Petroleum in the Marine Environment

*Workshop on Inputs, Fates, and the
Effects of Petroleum in the Marine Environment
May 21–25, 1973   Airlie House, Airlie, Virginia
Held under the Auspices of the*
OCEAN AFFAIRS BOARD
Commission on Natural Resources, National Research Council

NATIONAL ACADEMY OF SCIENCES / WASHINGTON, D.C. / 1975

NOTICE: The project that is the subject of this report was approved by the Governing Board of the National Research Council, acting in behalf of the National Academy of Sciences. Such approval reflects the Board's judgment that the project is of national importance and appropriate with respect to both the purposes and resources of the National Research Council.

The members of the committee selected to undertake this project and prepare this report were chosen for recognized scholarly competence and with due consideration for the balance of disciplines appropriate to the project. Responsibility for the detailed aspects of this report rests with that committee.

Each report issuing from a study committee of the National Research Council is reviewed by an independent group of qualified individuals according to procedures established and monitored by the Report Review Committee of the National Academy of Sciences. Distribution of the report is approved, by the President of the Academy, upon satisfactory completion of the review process.

International Standard Book Number 0-309-02311-4
Library of Congress Catalog Card Number 74-18572

Available from
Printing and Publishing Office, National Academy of Sciences
2101 Constitution Avenue, N.W., Washington, D.C. 20418

Printed in the United States of America

080918

Preface

During the last several years there has been an increasing concern over the possible effects of petroleum hydrocarbons in the marine environment and a number of studies have been conducted on this and related subjects. During 1972 the feeling developed in several circles that it was time for a comprehensive review of the state of knowledge in this area. Because of this concern, the Ocean Affairs Board of the National Research Council–National Academy of Sciences organized a workshop on inputs, fates, and effects of petroleum in the marine environment, which was supported by the Environmental Protection Agency, the U.S. Coast Guard, the Office of Naval Research, The Rockefeller Foundation, and the American Chemical Society. This report is an outgrowth of that workshop.

In the fall of 1972, an *ad hoc* Steering Committee was appointed (John M. Hunt, Woods Hole Oceanographic Institution; Clayton McAuliffe, Chevron Oil Field Research Company; Alan Walton, Bedford Institute of Oceanography; and Reino Kallio, University of Illinois) to organize the workshop and propose a list of possible participants. At that time, it was agreed to investigate the various sources of inputs of petroleum into the oceans and their amounts, to evaluate critically the methods for the chemical analysis of hydrocarbons and some of the appropriate techniques of biological study, to examine the fates of petroleum, and to study critically the information on the effects of oceanic pollution by petroleum. The workshop was primarily concerned with the evaluation of existing information rather than the development of new research. It was also expected that the workshop would indicate gaps in our knowledge and provide directions for new investigations.

The final organization of the workshop consisted of a Chairman (E. Bright Wilson, Harvard University); a Vice Chairman (John M. Hunt, Woods Hole Oceanographic Institution); four Area Chairmen; and eight Panel Leaders.

The participants—approximately 60 scientists and engineers from U.S. and foreign academic, governmental, and industrial organizations—were assigned to the panels. Panel members prepared position papers relative to their fields of expertise. Copies of these papers, which were circulated and available to all members prior to the meeting, constitute the basis of this workshop report. A limited number of the combined set of papers are available through the Ocean Affairs Board, National Research Council, as *Background Papers for a Workshop on Inputs, Fates, and Effects of Pe-*

troleum in the Marine Environment, Volumes I and II (1973). When this supply is exhausted, these papers will be available through the U.S. Department of Commerce's National Technical Information Service, publication number 163062.

During the Airlie House meeting, panels of experts in each area met to review the background papers and, based on their own general knowledge, to prepare the first draft of this report. General sessions were also held to provide the benefit of criticism from the entire group for each panel report.

The present document, although based on the workshop panel reports, is the responsibility of the General Chairman, the Vice Chairman, and the four Area Chairpersons. Changes from the original background and panel reports were not extensive and occurred mainly in areas where additional facts could be taken into consideration.

In all controversial projects such as this one, it is important to assemble experts from a broad spectrum of experiences and backgrounds. This we attempted to do. The text indicates that there is a paucity of firm data on petroleum in the sea, so that many of the conclusions are based on judgment rather than fact. Judgments inevitably are influenced by personal bias. Nevertheless, we are convinced of the sincerity of the panels, the overall balance of the workshop membership, and the efforts made to minimize the uncertainties introduced by these judgment factors. In the future, the uncertainty regarding petroleum in the marine environment can only be narrowed by the accumulation of more firm information. Until then we must be content with figures of less than complete certainty. It is to be hoped that those who quote or use data from this report will bear this in mind and not endow them with more authority than we claim.

It is difficult to pay sufficiently strong tribute to the effort, the devotion to the task, and the sincerity displayed by the members of this workshop. They met mornings, afternoons, and evenings daily. Divergences of viewpoints were discussed and mostly reconciled without rancor, despite the importance of many questions and the difficulty of obtaining reliable data.

E. BRIGHT WILSON
JOHN M. HUNT

Workshop Staff

Steering Committee

E. BRIGHT WILSON, *Chairman*
JOHN M. HUNT, *Vice Chairman*

JAMES N. BUTLER
CLAYTON D. MCAULIFFE
RUTH PATRICK
ERMAN A. PEARSON

Area Chairmen

ERMAN A. PEARSON, Inputs
CLAYTON D. MCAULIFFE, Chemical and Biological Methods
JAMES NEWTON BUTLER, Fates
RUTH PATRICK, Effects

Panel Leaders

CHARLES BATES, Inputs
FRED T. WEISS, Chemical Analysis
GILLES LA ROCHE, Biological Techniques
W. D. GARRETT, Physical Fates
JOHN M. TEAL, Biological Fates
JOSEPH CONNELL, Chronic Effects
BOSTWICK H. KETCHUM, Acute Effects
CYRUS LEVINTHAL, Health Effects

Ocean Affairs Board

RICHARD C. VETTER, *Executive Secretary*
MARY HOPE M. KATSOUROS, *Staff Officer*

Participants

E. R. Adlard, Shell Research Limited
Donald Ahearn, Georgia State University
Jack Anderson, Texas A&M
Charles C. Bates, U.S. Coast Guard
Donald S. Boesch, Virginia Institute of Marine Science
Ralph Brown, Esso Research and Engineering
K. T. Brummage, British Petroleum Trading Ltd., U. K.
James N. Butler, Harvard University
Donald Button, University of Alaska
D. B. Charter, U.S. Coast Guard
Ralph Churchill, University of California at Berkeley
R. B. Clark, University of Newcastle-upon-Tyne, England
R. C. Clark, NOAA Northwest Fisheries Center
H. D. Van Cleave, Environmental Protection Agency
Joseph H. Connell, University of California at Santa Barbara
Ralph Churchill, University of California at Berkeley
W. J. Cretney, Department of the Environment, British Columbia, Canada
Robert A. Duce, University of Rhode Island
John Farrington, Woods Hole Oceanographic Institution
Milton Feldman, Environmental Protection Agency
Donald L. Feuerstein, Engineering Science Inc.
G. D. Floodgate, University College, Bangor, Wales
William Garrett, Naval Research Lab
Conrad Gebelein, Bermuda Biological Station
Harry Gelboin, National Cancer Institute
Donald C. Gordon, Bedford Institute of Oceanography
Fred Grassle, Woods Hole Oceanographic Institute

Wilfred Gunkel, Biologische Anstalt Helgoland
Andres Hallhagen, IVLAB, Sweden
D. P. Hoult, Massachusetts Institute of Technology
John M. Hunt, Woods Hole Oceanographic Institution
Lawson Hunter, Department of Consumer Affairs, Ottawa, Canada
Mary Hope M. Katsouras, National Academy of Sciences
Bostwick Ketchum, Woods Hole Oceanographic Institution
Bruce Koons, Esso Production Research Co.
Pieter Korringa, Rijksinstituut voor Visserijonderzoek
Gilles La Roche, Consortium of Universities for Water Research, Montreal, Quebec, Canada
R. F. Lee, Scripps Institution of Oceanography
Cyrus Levinthal, Columbia University
Eric M. Levy, Bedford Institute of Oceanography
Clayton D. McAuliffe, Chevron Oil Field Research Co.
Edward Mertens, Chevron Research
Steven Moore, Massachusetts Institute of Technology
Wheeler J. North, California Institute of Technology
Christopher P. Onuf, University of California at Santa Barbara
Ruth Patrick, Academy of Natural Sciences
Erman Pearson, University of California at Berkeley
J. D. Porricelli, U.S. Coast Guard
Eugene Sawicki, Environmental Protection Agency
Lyle S. St. Amant, Louisiana Wildlife and Fisheries Commission
Howard Sanders, Woods Hole Oceanographic Institution
David Schultz, University of Rhode Island

James E. Stewart, Halifax Laboratory of Fisheries, Research Board of Canada
Philip N. Storrs, Engineering Science, Inc.
Michael Suess, World Health Organization
James Sullivan, Center for Science in the Public Interest
John Teal, Woods Hole Oceanographic Institute
Bruce Tripp, Woods Hole Oceanographic Institution
Richard C. Vetter, National Academy of Sciences
F. T. Weiss, Shell Development Co.
E. Bright Wilson, Harvard University
Richard D. Wilson, Exxon Production and Research Co.

Contents

1 INPUTS ... 1

Summary of Estimated Worldwide Inputs ... 1
Rationale and Basis for Estimates of PHC Release to the Ocean ... 2
    Natural Seeps, Offshore Production, Transportation Losses, Coastal Refineries, Atmosphere, Coastal Municipal Wastes and Coastal Nonrefinery Industrial Wastes, Urban Runoff, River Runoff
Future Trends ... 12
    Seeps, Offshore Production, Transportation Losses, Coastal Refineries, Atmosphere, Municipal and Industrial (Nonrefinery), Urban Storm Runoff, River Runoff
Conclusions ... 14
    Atmospheric Data, River Runoff, Natural Seeps, Sampling and Analysis
Recommendations ... 15

APPENDIX: Characterization of U.S. Statistics for Petroleum Accidentally Released into the Sea ... 15

REFERENCES ... 17

2 ANALYTICAL METHODS ... 19

CHEMICAL ... 19
General Considerations ... 19
    Hydrocarbon Types in Crude Oils, Hydrocarbon Types in Refined Petroleum Products, Biogenic Hydrocarbons, Differentiation of Petroleum Hydrocarbons from Biogenic Hydrocarbons
Analysis of Oils ... 21
    Sample Collection, Sample Preservation, Preparation of Oil Samples for Analysis, Analytical Methods
Hydrocarbons in Water ... 22
    Sample Collection and Preservation, Analytical Methods for C_1–C_{10} Hydrocarbons, Analytical Methods for C_{11} Plus Hydrocarbons
Hydrocarbons in Biological Materials ... 28
    Sample Collection and Preservation, Extraction and Saponification, Separation of Hydrocarbons from Lipids, Analytical Methods

Hydrocarbon Analysis of Sediments 30
    Sample Collection and Preservation, Analytical Methods
Summary and Recommendations 31

BIOLOGICAL 32

Microbial Biodegradation 32
    Population Enumeration
Metabolic Activity 33
Hydrocarbon Bio-Uptake 33
Algal Responses 33
    Phytoplankton, Attached Macroscopic Algae, Benthic Microflora
Sedentary Invertebrate Populations 34
    Sampling Benthos, Benthic Analysis
Intertidal Studies 35
Experimental Toxicity on Invertebrates and Fish 36
    Acute Bioassays, Behavioral Studies, Other Biological Changes
Summary and Recommendations 37

REFERENCES 38

3 FATES 42

PHYSICAL AND CHEMICAL 42

Physical and Chemical Characteristics of Petroleum 42
Processes 43
    Spreading, Evaporation, Solution, Emulsification, Direct Sea-Air Exchange, Photochemical Oxidation, Tar Lump Formation, Sedimentation
Summary: The Life History of a Spill 51
Amounts of Hydrocarbons in the Marine Environment 52
    Atmosphere, Sea Surface Microlayer, Pelagic Tar Lumps, Water Column, Sediments
Conclusions 55
Recommendations 55

BIOLOGICAL 58

Microbial Degradation 58
Uptake by Organisms 60
Metabolism 64
Storage 65
Discharge 65
Food Web Magnification 66
Conclusions 66
Recommendations 67

REFERENCES 67

4 EFFECTS 73

Factors Influencing the Biological Impact of Oil Spills 83
    Oil Dosage, Oil Type, Oceanographic Conditions, Meteorological Conditions, Turbidity, Season, Biota Type, Methods of Oil Spill Cleanup
Effects on Various Types of Habitats 84
    Intertidal Areas, Rocky Substrates, Sandy Substrates, Marshland, Subtidal Substrates
Effects on Aquatic Organisms 85
    Productivity, Primary Producers, Detritus Feeders, Reproduction, Growth, Respiration, Behavior, Histology, Evaluation of Available Information

Effects on Aquatic Populations and Communities 88
    Plankton, Subtidal Organisms, Fisheries, Persistence of Effects on Communities, Recovery of Communities, Community Diversity
Effects on Seabirds 92
    Chronic Pollution, Population Changes, Remedial Measures
Need for Evaluating Biological Effects of Oil in the Marine Environment 94
    Types of Biological Studies, Questions that Need Answers
Direct Impact on Humans 95
    Tar and Oil Pollution of Beaches, Possible Human Health Effects
Conclusions 98

REFERENCES 100

5 CONCLUSIONS 104

1 Inputs

There are major problems in estimating the input and the flux of petroleum hydrocarbons (PHC) in the environment both generally and with respect to the oceans in particular. There is a paucity of reliable analytical data on the concentrations of PHC in the volume fluxes of air, liquids, and solids. This problem is compounded further by the absence of generally accepted standard methods for PHC analyses. At best, the results represent estimates obtained by analytical methods ranging from column separation and quantitative determination by infrared analysis (Lindgren, 1957) to a simple solvent extraction (n-hexane)/gravimetric analysis, the conventional wastewater method for oil and grease (American Public Health Association, 1971). The latter method has been supplemented on occasion by a saponification procedure to crudely estimate the PHC fraction. Insufficient data are reported in the literature to differentiate between levels of biogenic and petroleum hydrocarbons in most of the sources, or to identify or quantify the various components making up the general class termed *petroleum hydrocarbons*. In addition, problems exist even where data are available; for example, volumes of PHC lost through industrial practices and accidents, such as spills, in the territorial and high seas are probably reported on the low side.

To estimate as accurately as possible the flux of PHCs to the environment from all significant sources, we assembled a panel of experts from various professional disciplines, including geologists, geochemists, and petroleum, chemical, and sanitary engineers from the United States, Sweden, and the United Kingdom. Best estimates were developed for each significant petroleum source based on the limited availability of reliable data and modified by judgments derived from extensive experience with each source. The panel had similar estimates made by other groups available during their deliberation. Among these were the 1970 Study of Critical Environmental Problems (SCEP), the technical studies prepared in early 1973 for the tenth session of the Subcommittee on Marine Pollution of the Intergovernmental Maritime Consultative Organization, and the U.S. Coast Guard's updated (1973) quantitative estimates of petroleum being introduced into the oceans.

SUMMARY OF ESTIMATED WORLDWIDE INPUTS

The following summary compiles estimated worldwide PHC inputs to the oceans, encompassing all petroleum sources considered significant by the panel members. These range in types from extremely diffuse sources to occasional major point sources of variable location, such as tanker accidents. Clearly, the importance or significance of a particular source depends not only on its relative size but also on the nature of the source and the scope and degree of possible effects.

The time frame for which these compiled estimates apply is diffuse. The key figures pertaining to petroleum produced and in marine transit are for 1971 (British Petroleum Company, Ltd., 1971). Spill data for the United States were derived from the U.S. Department of the Interior, which uses 1971–1972 for establishing loss rates on the outer continental shelf, and the U.S. Coast Guard, which uses 1970–1971. The time frame for sanitary engineering measurements for waste petroleum from shoreside installations was from 1964 to 1972. The basic atmospheric emission values were measured during 1968.

In compiling such a summary, it is difficult to avoid repeating sources of petroleum hydrocarbons. For example, municipal and industrial wastes and refinery inputs were estimated separately for coastal zones and for inland areas, although the inputs from inland areas made up some part of the river runoff inputs. However, every effort was made to ensure that there is no overlap or double accounting of sources of petroleum.

Because most of the world's crude petroleum is now obtained from countries other than those in which the petroleum is refined and consumed, most of this petroleum is transported thousands of miles by ships and pipelines. The world production pattern is shown in Table 1-1. Table 1-2 shows the world consumption pattern, and Table 1-3 shows the resulting total inter-area oil movement. Figure 1-1 shows the international flow of petroleum in accordance with data from the U.S. Department of the Interior. Table 1-4 shows imports and exports for crude oil and its by-products for 1971.

The rate at which crude petroleum and its by-products are actually entering the ocean is, of course, impossible to determine with complete accuracy. The introduction rate given in Table 1-5 is based on our evaluation of the best available data. Table 1-6 shows values for the introduction of PHCs into the ocean as tabulated by the SCEP study of the Massachusetts Institute of Technology in 1970 and by the U.S. Coast Guard in mid-1973. With the exception of atmospheric "rain-out" values, our values were somewhat higher than those presented in the above-mentioned publications. We believe that this difference is largely due to the use of a more comprehensive data base and by subsequent experience with the assumptions and calculations used to arrive at a final global value for any particular type of input. These assumptions and calculations are described below.

RATIONALE AND BASIS FOR ESTIMATES OF PHC RELEASE TO THE OCEAN

NATURAL SEEPS

The direct input of oil from natural seeps into the marine environment is estimated to be 0.6 million metric tons per annum (mta), with a range from 0.2 to 1.0 mta. This estimate is based on geological and geochemical criteria, as described by Wilson *et al.* (1973).

Figure 1-2 shows the 190 known significant world-

TABLE 1-1 World Oil Production, 1971

Country (or Area)	Million Tons
North America	
United States	
Crude oil	473.2
Natural gas liquids	60.1
	533.3
Canada	77.1
Mexico	24.9
TOTAL	635.3
Caribbean	
Venezuela	187.3
Colombia	11.6
Trinidad	6.8
TOTAL	205.7
South America	
Argentina	21.9
Brazil	8.3
Other	6.6
TOTAL	36.8
TOTAL WESTERN HEMISPHERE	877.8
Western Europe	
France	1.9
W. Germany	7.4
Austria	2.5
Turkey	3.5
Other	7.1
TOTAL	22.4
Middle East	
Iran	226.2
Iraq	83.4
Kuwait	147.1
Neutral Zone	28.3
Qatar	20.5
Saudi Arabia	223.4
Abu Dhabi	44.9
Oman	14.4
Other	15.5
TOTAL	803.7
Africa	
Algeria	35.9
Libya	132.9
Other North Africa	25.7
Nigeria	74.1
Other West Africa	11.9
TOTAL	280.5
Southeast Asia	
Indonesia	43.9
Other	11.9
TOTAL	55.8
Soviet Union	372.0
Eastern Europe and China	42.0
Other Eastern Hemisphere	24.2
TOTAL EASTERN HEMISPHERE	1,600.6
WORLD (excluding the Soviet Union, Eastern Europe, and China)	2,064.4
WORLD TOTAL	2,478.4

TABLE 1-2 World Oil Consumption, 1971

Country (or Area)	Million Tons
United States	715
Canada	77
Mexico	26
Caribbean	56
South America	66
TOTAL WESTERN HEMISPHERE	940
Belgium and Luxembourg	29
Netherlands	36
France	103
W. Germany	133
Italy	92
United Kingdom	103
Scandinavia	54
Spain	27
Other Western Europe	75
TOTAL WESTERN EUROPE	652
Middle East	54
Africa	45
South Asia	29
Southeast Asia	62
Japan	220
Australasia	30
Soviet Union, Eastern Europe, and China	364
Eastern Europe, Asia, and Africa	804
TOTAL EASTERN HEMISPHERE	1,456
WORLD (excluding the Soviet Union, Eastern Europe, and China)	2,032
WORLD TOTAL	2,396

wide marine seeps; however, little has been reported in the literature regarding the nature and magnitude of these seeps. The similarity of geologic and geochemical factors that affect oil seepage on land and in areas adjacent to the continents was a major consideration in assessing the input from offshore marine seeps. Based on a study of these factors, Wilson estimated that the world's continental margins were subdivided into areas of high, moderate, and low seepage potential. His seepage rates—given as 100, 3, and 0.1 barrels/day/1,000 sq mi for high, moderate, and low seepage, respectively—were then applied to each area based on the geological nature of the region. Calculations based on the assigned rates then yielded an estimate of total annual worldwide seepage from the offshore areas of 0.6 mta.

Estimates in this range have been criticized by Blumer (1972) as too high on the grounds that such a rate would have depleted all the oil in place in offshore reservoirs. A review of this question with petroleum geologists results in agreement that the problem of estimating natural seepage is extremely complex. Seeps tend to occur in tectonically active areas containing petroleum source beds. Since most petroleum is generated at depths generally greater than 3,000 ft, there is not much loss to the surface from either source or reservoir beds until the geological sections are uplifted and eroded or until the updip extensions are exposed. It has been estimated that 50 to 100 times as much oil has been lost to the environment as now exists in reservoirs. These estimates are based on major seeps such as the Athabasca Tar Belt of the Western Canada Basin. However, most of this oil has been lost to the atmosphere or to rivers cutting into the eroded basins on the continents. The oceanic problem is somewhat different, although the geological factors are similar. There is much less organic matter in sediments seaward of the continental rise, compared with basins where seas have invaded the continents. This is not unexpected, since the open ocean is largely a biological desert except for the upwelling areas. Also, the small amount of organic matter that does accumulate in the deep ocean is readily oxidized since rates of deposition are very slow.

All of this makes it questionable whether seepage rates in the marine environment have been comparable with those on the continents in the geologic past. If not, then the average seepage over geologic time should be less than the best proposed estimate in this report.

OFFSHORE PRODUCTION

The worldwide input of oil to the ocean from offshore drilling and production operations is estimated at 0.08 mta. Of this total, 0.02 mta is estimated to be lost through minor spills (50 barrels or less) and through discharges of oil field brines during normal drilling and producing operations. The remaining 0.06 mta is lost during major accidents, spills of greater than 50 barrels that result from blowouts, rupture of gathering lines, and similar unpredictable occurrences. These estimates are based on data obtained from United States operations on the Gulf Coast (U.S. Department of the Interior, 1972a, 1973) and have been extrapolated worldwide with appropriate allowances made for differences in operating procedures and standards outside the United States. The procedures and data used are summarized below.

Losses during Normal Drilling and Producing Operations

According to the U.S. Department of the Interior (1973), a total of 836 barrels of oil was lost through minor offshore operation spills in the Gulf of Mexico during the first 9 months of 1972, which amounts to 3.1 barrels of oil per day. Assuming that 1972 production at least equaled the 1971 production of 1.2 million barrels per day (McCaslin, 1972), the amount of oil lost to the Gulf via minor spills was 2.6×10^{-6}

TABLE 1-3 Interarea Total Oil Movements, 1971

Million Tons of Oil From	United States	Canada	Other Western Hemisphere	Western Europe	Africa	Southeast Asia	Japan	Australasia	Other Eastern Hemisphere	Destination Not Known	Total Exports
United States	–	1.50	4.00	3.75	0.25	0.25	2.00	0.25	0.25	–	12.25
Canada	39.25	–	–	–	–	–	–	–	–	–	39.25
Caribbean	113.50	21.00	2.00	27.25	0.25	–	0.75	–	–	–	164.75
Other western hemisphere	1.00	–	–	–	–	–	–	–	–	–	1.00
Western Europe	7.50	–	–	–	2.25	–	0.50	–	1.25	4.50	16.00
Middle East	20.00	13.50	22.25	378.75	25.25	48.50	191.75	15.50	20.25	19.50	755.50
North Africa	4.50	–	4.50	158.25	0.25	–	1.00	–	10.50	7.00	185.75
West Africa	5.50	2.50	6.25	55.50	–	–	2.00	–	–	7.50	79.25
Southeast Asia	6.75	–	0.50	0.25	0.25	–	31.25	3.00	–	–	42.00
Soviet Union, Eastern Europe	0.25	–	7.00	43.75	3.00	1.50	1.50	–	0.50	–	56.00
Other eastern hemisphere	0.50	–	–	0.50	0.25	–	0.25	–	–	–	3.00
TOTAL IMPORTS	198.75	38.50	46.50	668.00	31.75	50.25	231.00	18.75	32.75	38.50	1,354.75

TABLE 1-4 Imports and Exports of Crude Oil and Products, 1971

| | Million Tons of Oil | | | |
| | Imports | | Exports | |
Country (or Area)	Crude	Products	Crude	Products
United States	83.25	115.50	–	12.25
Canada	32.50	6.00	35.50	3.75
Caribbean	11.00	3.75	51.75	113.00
Other western hemisphere	28.00	3.75	–	1.00
Western Europe	633.25	34.75	0.75	15.25
Middle East	6.25	–	691.25	64.25
North Africa	4.00	4.00	183.50	2.25
West Africa	0.25	1.00	78.75	0.50
East and South Africa, South Asia	36.50	6.50	–	0.50
Southeast Asia	35.25	15.00	32.25	9.75
Japan	189.50	41.50	–	1.00
Australasia	15.00	3.75	0.75	0.75
Soviet Union, Eastern Europe, China	5.25	0.75	28.50	27.50
Destination not known[a]	23.00	15.50	–	–
TOTAL	1,103.00	251.75	1,103.00	251.75

[a] Includes changes in stock and quantities in transit, transit losses, minor movements not otherwise shown, military use, etc.

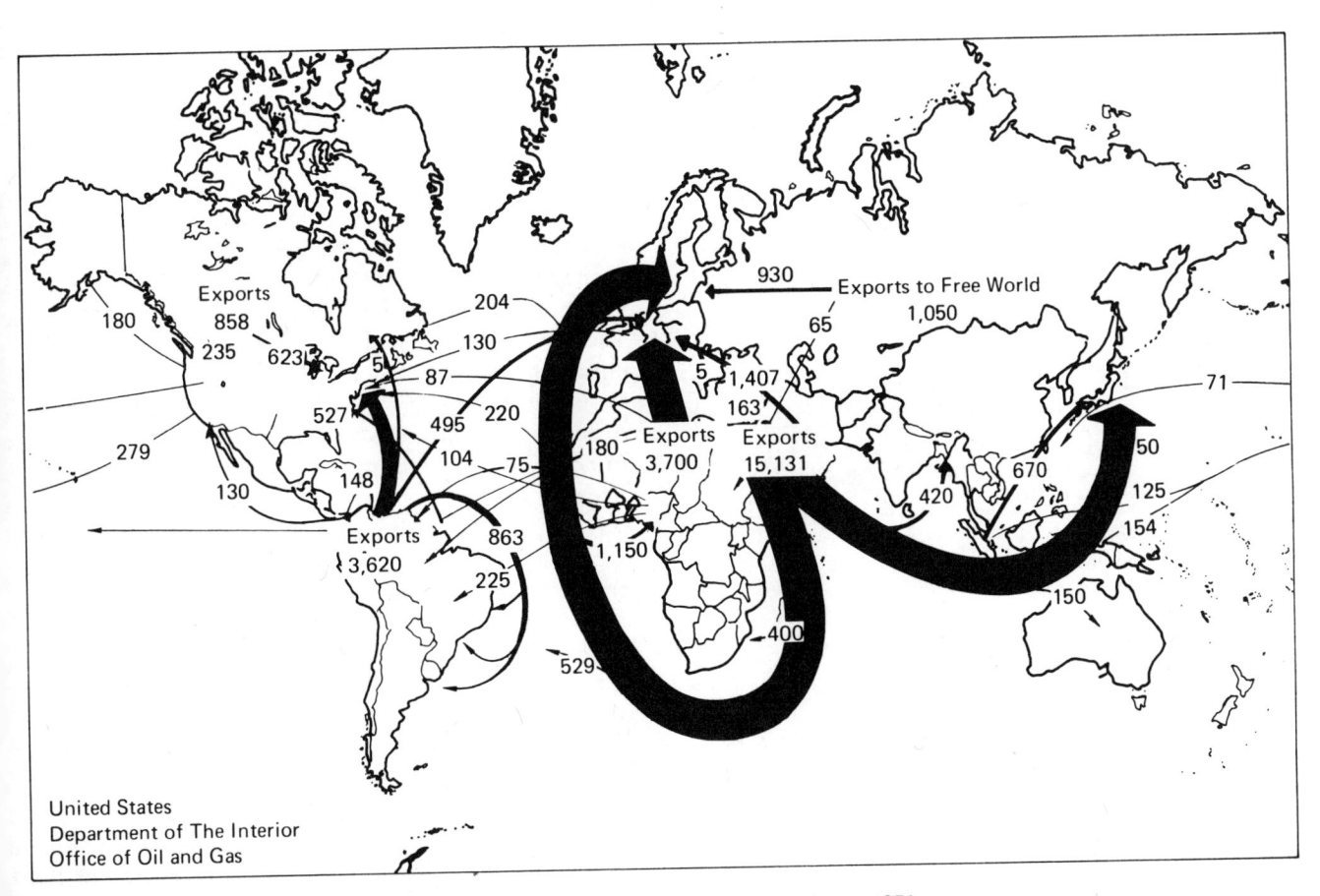

FIGURE 1-1 International flow of petroleum, 1971.

TABLE 1-5 Budget of Petroleum Hydrocarbons Introduced into the Oceans

Source	Best Estimate	Probable Range	Reference
Natural seeps	0.6	0.2–1.0	Wilson et al. (1973)
Offshore production	0.08	0.08–0.15	Wilson et al. (1973)
Transportation			
LOT tankers	0.31	0.15–0.4	Results of workshop
Non-LOT tankers	0.77	0.65–1.0	panel deliberations
Dry docking	0.25	0.2–0.3	
Terminal operations	0.003	0.0015–0.005	
Bilges bunkering	0.5	0.4–0.7	
Tanker accidents	0.2	0.12–0.25	
Nontanker accidents	0.1	0.02–0.15	
Coastal refineries	0.2	0.2–0.3	Brummage (1973a)
Atmosphere	0.6	0.4–0.8	Feuerstein (1973)
Coastal municipal wastes	0.3	–	Storrs (1973)
Coastal, Nonrefining, industrial wastes	0.3	–	Storrs (1973)
Urban runoff	0.3	0.1–0.5	Storrs (1973), Hallhagen (1973)
River runoff	1.6	–	Storrs (1973), Hallhagen (1973)
TOTAL	6.113		

[a] mta, million metric tons.

barrels per barrel of oil produced. This loss factor should be representative of other U.S. offshore operations. In this case, the total loss due to minor spills from U.S. offshore operations is 1,500 barrels per year (based on 1971 U.S. offshore production of 618 million barrels, McCaslin, 1972).

In some other parts of the world, quantitative reports suggest that losses due to minor spills may average 10 times greater than those that occur in U.S. waters. Accordingly, the loss due to minor offshore spills outside the United States is estimated to be 62,000 barrels per year (based on 1971 foreign offshore production of 2,580 million barrels per year, McCaslin, 1972). Thus, the total worldwide loss of oil through minor spills during normal offshore operations is estimated at 63,500 barrels or approximately 0.01 mta.

The estimate of oil loss via field brine discharges is obtained in a similar manner. Brines that are produced along with the oil and gas are usually discharged into the sea after passing through an oil-water separator. These brines still contain small amounts of oil. Under present federal regulations, this amount of oil cannot exceed 50 parts per million (ppm) of produced brine. According to the U.S. Department of the Interior (1972b), 7.3 barrels of waste oil per day were discharged into the Gulf of Mexico, along with 180,000 barrels per day of brine, resulting in an average oil content of 41 ppm. This occurred during the daily production in the Gulf of Mexico of 1.2 million barrels of oil during 1971. It is equivalent to a loss of 6.0 barrels per million barrels of oil produced.

Depending on the separator used, the oil content of treated brines in other parts of the world may be up to four times higher than the level allowed in the Gulf of Mexico at present. While worldwide data on volumes of produced brines are not available, it is not likely that economic self-interest will allow higher brine percentages from producing wells in other parts of the

TABLE 1-6 Comparison of Estimates for Petroleum Hydrocarbons Annually Entering the Ocean, circa 1969–1971

Source	MIT SCEP Report (1970)	USCG Impact Statement (1973)	NAS Workshop (1973)
Marine transportation	1.13	1.72	2.133
Offshore oil production	0.20	0.12	0.08
Coastal oil refineries	0.30	–	0.2
Industrial waste	–	1.98	0.3
Municipal waste	0.45	–	0.3
Urban runoff	–	–	0.3
River runoff[a]	–	–	1.6
SUBTOTAL	2.08	3.82	4.913
Natural seeps	?	?	0.6
Atmospheric rainout	9.0[b]	?	0.6
TOTAL	11.08	?	6.113

[a] PHC input from recreational boating assumed to be incorporated in the river runoff value.
[b] Based upon assumed 10 percent return from the atmosphere.

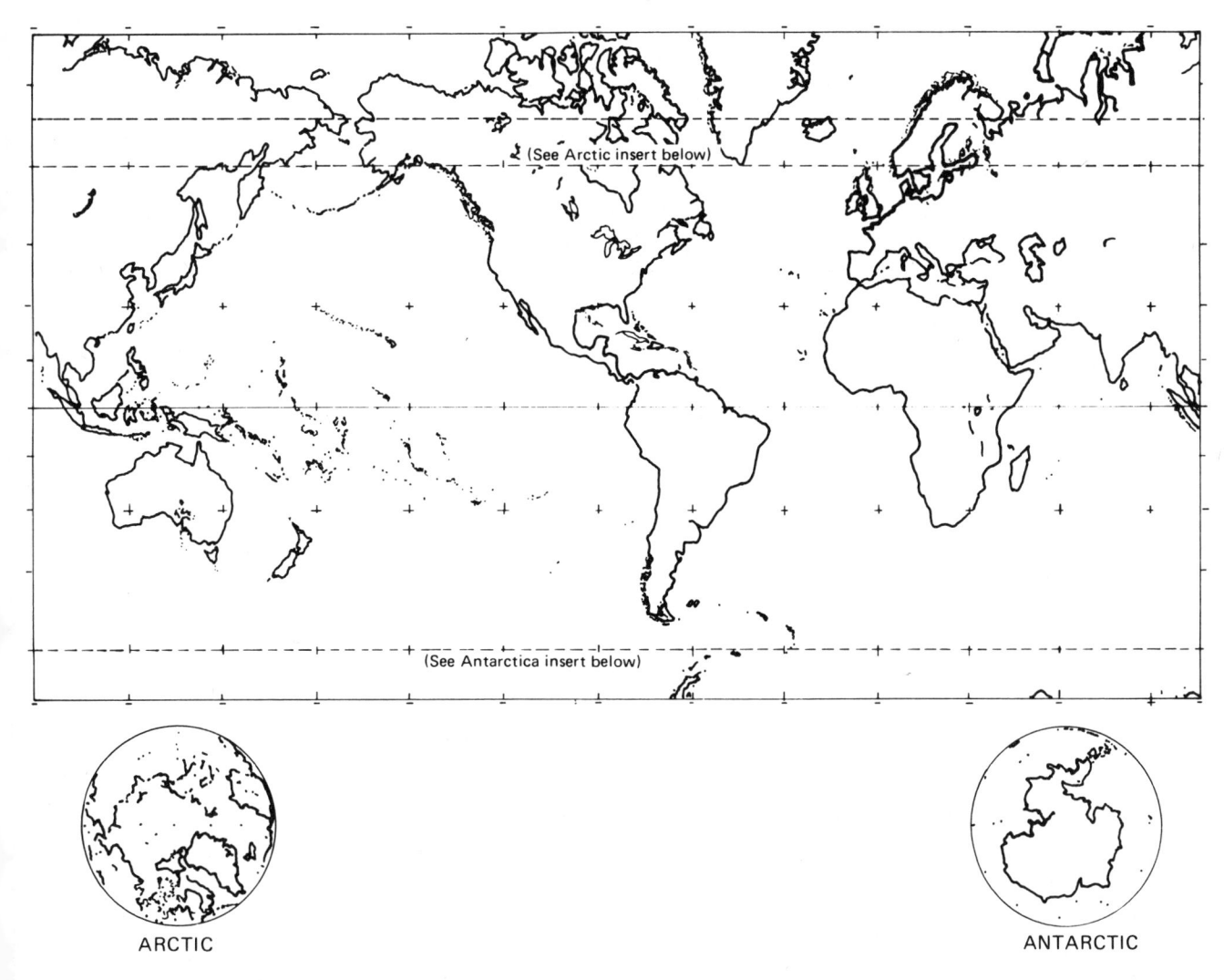

FIGURE 1-2 Location of reported submarine seeps.

world than produced in U.S. waters. Therefore, as an upper limit, it can be assumed that the same proportion of brine and oil will be produced worldwide. Application of the above loss factor to the 1971 annual offshore production data previously cited results in a worldwide estimate of oil entering the sea with produced brine at 64,600 barrels per year, or 0.01 mta. Thus, about 0.02 mta of oil are lost to the marine environment during offshore drilling and producing operations—0.01 mta from minor spills and 0.01 mta entrained in the produced brines.

Accidents Causing Major Spills of More Than 50 Barrels

Estimates of losses due to unpredictable offshore accidents resulting in spills of greater than 50 barrels of oil are based on U.S. Department of the Interior (1973) records of industry experience in the Gulf of Mexico from 1964 to 1971. These data yield an average annual loss of 0.014 percent of production, ranging from 0.0003 to 0.7 percent. It can be assumed that in the future, accidental losses should not exceed the experience of this 8-year period (U.S. Department of the Interior, 1972a). Also, the U.S. loss of petroleum through this source should be similar with other nations because the same safety precautions, operating equipment, and procedures are used by industry on a worldwide basis. Accordingly, by applying the 0.014 percent accidental loss factor to 1971 worldwide offshore production of 3,200 million barrels of oil (McCaslin, 1972), it is estimated that unpredictable episodic accidents may result in annual losses of about 450,000 barrels of oil or 0.06 mta.

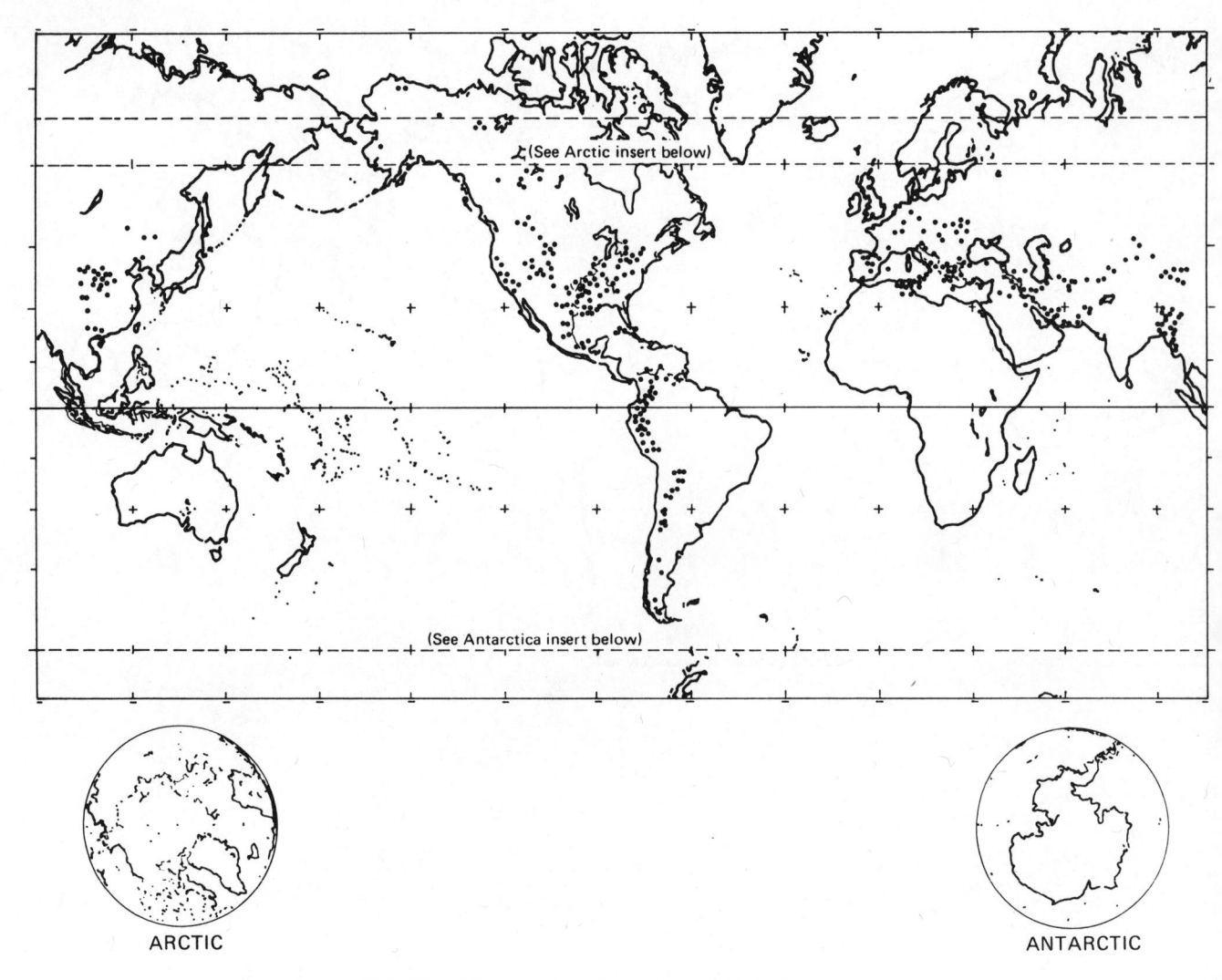

FIGURE 1-3 Location of some reported onshore seeps.

TRANSPORTATION LOSSES

Tankers

Total world oil production in 1971 was approximately 2,400 million tons (British Petroleum Company, Ltd., 1971), of which 1,355 million tons (including 1,100 million tons of crude oil) was transported by sea. A schematic diagram of this flow pattern prepared by the U.S. Department of the Interior is shown in Figure 1-1. Out of a total of more than 50,000 merchant ships afloat, 6,000 ships, with a carrying capacity of 180 million dead weight tons, were used to transport this oil. In addition, as shown in Table 1-4, prepared by British Petroleum Company, Ltd. (1971), another 250 million tons of refined petroleum products moved across national borders, predominantly by ship.

It is normal practice to wash the cargo tanks of tankers with seawater, and, in the past, crude oil tankers have discharged these washings overboard. The oil industry has therefore introduced a procedure, known as the Load On Top (LOT), to minimize this loss. In this procedure, washings and oily water from ballast are retained on board the ship for settling to concentrate their oil content, which is then incorporated into the next shipment.

If no measures were taken to avoid operational oil discharges by tankers, the average amount discharged in a ballast voyage by a tanker would be about 0.35 percent of the carrying capacity of the ship. This percentage, though, can range from as low as 0.1 percent of carrying capacity for light refined products to as much as 1.5 percent for residential fuel oils. On the basis of a total crude oil transported by sea of 1,100 mta, the potential loss to the ocean from this source, if the LOT method were not in use, would be about 3.85 mta. At present, however, oil industry statistics indicate that about 80 percent of tankers use LOT; hence, the ships not using it are probably discharging about 0.77 mta to the sea.

This results in the potential saving from the use of LOT of 3.1 mta. However, even when LOT is used efficiently, there is a small loss of oil to the sea because the performance of a tanker depends on the concern and proficiency of tanker crews. Thus, in practical situations, efficiency, which could approach 99 percent, is considered to be in the actual range of 90 percent (Victory, 1973); thus, the loss from LOT tankers would be 0.31 mta. If one then adds the two, a grand total of 1.08 mta in losses from operational discharges of tankers is obtained. Further, there remains the question of whether the quantities of oil discharged during difficult wave conditions, short hauls such as in the Mediterranean Sea when LOT does not have time to become effective, and poor supervision might lower the average LOT effectiveness value below 90 percent.

Dry Docking

A further source of pollution arises in connection with the need for tankers to go into dry dock for inspection, maintenance, or refitting at periodic intervals. At present the frequency of dry docking is between 1 and 2 years, averaging about 18 months. Since all cargo ships must be clean and gas-free for this purpose, rigorous cleaning of the entire ship is required and the oil removed. This is estimated to yield a quantity for disposal of about 0.4 percent of carrying capacity (IMCO, 1972b). Frequently, tankers arrive at the dry dock having discharged the total washings at sea (often because reception facilities for the washings are not available ashore). The proportion arriving without washing has been estimated to be about 50 percent. For a world fleet of 180 million dead weight tons the annual loss to the sea prior to dry docking would then be about 0.25 mta.

Terminal Operation

Milford Haven (United Kingdom), which predominately handles large tankers, experienced a spillage rate of 0.00011 percent of throughput over a period of 9 years (Brummage, 1973a). This percentage is considered reasonably representative of a well-controlled port of this type. The equivalent ratio for Portland, Maine, during 1972 was 0.00022 percent. Application of the Milford Haven spilling ratio to world oil marine movements of 1,355 mta and—remembering that this tonnage is handled twice, once when loaded and once when discharged—leads to a loss of 0.003 mta.

An important variable is the number of ship/shore transfers; these will be greater in many ports handling shipping smaller than very large crude carriers (VLCC) and be engaged in more local trade. A further factor, particularly in the United States, is lightering of tankers with the resultant large number of short haul trips by smaller tankers. Moreover, supervision is not uniformly strict throughout the world. Taking these factors into account, Porricelli et al. (1971) reported a spillage loss of 0.007 mta in 1971. This probably ranges between 0.0015 and 0.005 mta, with a preferred value of 0.003 mta.

Bilges and Bunkering

The rate of loss due to this source is difficult to support by means of measured data. Estimates seem usually to be based on an assumed rate of loss per ship, with reported total loss ranging from 0.02 mta (Holdsworth, 1971) to 0.7 mta (Porricelli et al., 1971). We estimated 0.5 mta (0.4–0.7 mta), bearing in mind all the inaccuracies involved and considering all sources including leaks and bunkering. This is equivalent to an average loss per ship of about 10 tons per annum.

Tanker Accidents

Several authors (Holdsworth, 1971; Porricelli et al., 1971; Porricelli and Keith, 1973) have reviewed the available statistics concerning losses due to tanker accidents, covering various periods and different ranges of ship size, and found the results to range from 0.05 to 0.25 mta. There are major variations in total spillage from year to year depending on the occurrence of major accidents. The most recent and complete study (Porricelli and Keith, 1973) strongly suggests that the rate of spillage is at the upper end of the range. A probable value of 0.2 mta was therefore adopted, with a possible range between 0.12 and 0.25 mta.

Nontanker Accidents

The data on these accidents are scanty. There are about nine times as many nontankers as tankers, but their average size is much smaller; also, the only oil normally carried in them in bulk is bunker fuel. Reported estimates for the annual rate of loss range from 0.02 mta (Holdsworth, 1971) to 0.25 mta (Porricelli et al., 1971). We consider the latter to be excessive. A more likely value is 0.10 mta, with a range between 0.02 and 0.15 mta.

COASTAL REFINERIES

Available estimates for the quantity of dispersed oil discharged in the water effluent from seaboard refineries are 0.20 mta (Holdsworth, 1971) and 0.30 mta (SCEP, 1970).

An unrealistically high estimate can be made by assuming that once-through water cooling is used throughout, so that the water usage is about 3,000 m³/h per million tons/annum refining capacity (Blokker, 1971). If gravity separation/dissolved air flotation is used, the oil content of the effluent will be about 20 ppm. Taking a seaboard refining capacity of about 1,000 mta, these figures lead to a loss of about 0.50 mta.

In fact, recirculating water cooling or air cooling is being used to an increasing extent, and there is also strong pressure to reduce the oil content below 20 ppm. Therefore, the published data are considered to represent the practical situation. We consider the preferred rate of loss at present to be about 0.20 mta, and it is expected to decrease even further within the next few years.

ATMOSPHERE

The total worldwide influx of petroleum-derived hydrocarbons to the atmosphere is estimated at 68 mta. This is based on a detailed inventory of hydrocarbon emissions by the National Air Pollution Control Administration (1968) and an appropriate scaling of these values to a world figure based on energy patterns. A similar detailed study of hydrocarbon emissions in Sweden, when scaled to world proportion yielded data that support this value (A. Hallhagen, personal communication, 1973). Also, there is good agreement with the value of 90 mta given by SCEP (1970), considering that some nonpetroleum hydrocarbons are included in that total.

Approximately two-thirds of the PHC entering the atmosphere, about 45 mta, emanate from transportation sources. Fuel combustion in stationary sources, industrial processes, and solvent and gasoline evaporation account for the remainder. PHC emissions from automobiles have been studied, particularly in relation to photochemical smog formation. These emissions are categorized as reactive and nonreactive, based on their relative rates of photochemical oxidation (Robinson and Robbins, 1968). Of the 45 mta of PHC entering the atmosphere from transportation sources, 22 mta (approximately one-half) are reactive. Each source was similarly divided into reactive and nonreactive categories based on a general knowledge of its chemical composition, resulting in reactive and nonreactive portions of 33 and 35 mta, respectively. In light of the relatively long mean residence times in the atmosphere and the distinct possibility of solid particulates serving as catalyst sites to increase reaction rates (Altschuller, 1966; Newell, 1971), it is assumed that all of the reactive fraction and 90 percent of the nonreactive fraction is eventually converted and does not return to the earth's surface as PHC. This yields approximately 4 mta PHC available to return to the earth's surface. This is admittedly a severe assumption, since the chemistry of PHC interactions has not been studied relative to final reaction products or absolute reaction rates. However, calculations based on average solid particulate concentrations in the atmosphere and on the assumption that PHC are associated (through absorption or similar phenomena), with particulates as a function of surface area, indicate that 4 mta is a reasonably valid value.

The amount of PHC entering the oceans as direct aerial fallout (either carried by rainfall or dry fallout) is estimated to be 10–20 percent of the 4 mta or 0.4–0.8 mta. This is based on the relative areas of land to oceans, on a consideration of global precipitation patterns, and on the fact that the atmospheric concentration of particulates is at least three orders of magnitude greater over land areas than over the ocean (Junge, 1963).

This estimate of influx of PHC to the ocean from the atmosphere is approximately one order of magnitude lower than that estimated by the SCEP study (1970). However, that study assumed that all of the PHC entering the atmosphere was available to be returned and that 10 percent returned as fallout to the oceans. Our estimate is based on assumptions concerning the attenuation and modification of influent PHC in the atmosphere such that the total amount available for fallout is less than that which enters the atmosphere. Additionally, it appears reasonable to assume that PHC are associated with atmospheric particulates that fall predominately on land. The major source of error arises from the paucity of data regarding the reaction kinetics of PHC, their resulting end products, and the PHC content of aerial particulate matter.

COASTAL MUNICIPAL WASTES AND COASTAL NONREFINERY INDUSTRIAL WASTES

A number of recent studies (Southern California Coastal Water Research Project, 1973; Storrs, 1973) conducted in major coastal cities have provided estimates of per capita contributions of oil and grease ranging from 6 to 27 g/day. Consideration of the varying contributions in relation to the degree of industrialization in each city suggested to us that a reasonable mean figure would be about 16 g/capita-day (g/cap-d). This is the mean contribution for the Southern California metropolitan complex (11 million persons) (Southern California Coastal Water Research Project, 1973) and the Island of Oahu (1 million persons) and falls within the 9–26 g/cap-d range estimated for Rio de Janeiro (about 3 million persons). We considered these areas to be representative of coastal areas worldwide.

Based on a number of determinations in California, it is assumed that about half the total oil and grease in municipal sewage is PHC, yielding a mean per capita contribution of petroleum hydrocarbons of 8 g/day. Additional analyses, specifically of PHC in sewage effluents, were generally similar to this estimate (Farrington and Quinn, 1973). Analysis of available data from the San Francisco and Los Angeles areas suggests that 4 g/cap-d result from municipal wastewaters (including gas stations, garages, commercial operations, etc.), with the remaining 4 g/cap-d due to a wide variety of industrial wastes discharged to the municipal sewer systems. These estimates of the industrial contribution of petroleum hydrocarbons do not include oils in refinery wastewaters that may or may not be connected to a municipal sewer system.

In the United States about 68 million people live in the coastal zone, so that the total United States discharge of oil in municipal (no major industries) wastewater would be 4 g/cap-d $\times$ 365 $\times$ 68 $\times$ 10^6 or about 0.1 mta. If it is further assumed that the worldwide input is proportional to the world: United States ratio of petroleum utilization (about 3:1), the world input from domestic wastewaters would be about 0.3 mta.

Since the unit inputs from industry (other than refineries) in coastal areas appears to be about the same as that from domestic wastewaters, it is assumed that coastal zone industry also discharges about 0.3 mta to the oceans.

Since industry in inland areas will discharge to rivers that reach the ocean, these industrial (and municipal) contributions must be accounted for in the river runoff calculations.

URBAN RUNOFF

Significant amounts of PHC are deposited in urban areas from a variety of sources, including oil heating systems, fallout, operation of automobiles, etc. Rainfall and runoff will inevitably wash these materials into storm drains and into receiving waters.

These inputs depend, to a great extent, on human activities and, therefore, should be related to population and petroleum utilization. In view of the data presented for PHC inputs from municipal and associated industrial waters, an estimate of PHC inputs from urban storm drainage might be developed as a fraction of the inputs from municipal and industrial wastewaters.

Analysis of annual emissions of oil and grease in municipal and industrial wastewaters and in storm drainage from several different areas in southern California indicated ratios of oil and grease from storm drainage to that in municipal and industrial wastewaters to be between about 0.1 and 0.6 (Southern California Coastal Water Research Project, 1973; Storrs, 1973). Oil and grease in storm drainage from the central Los Angeles area was about one-third (0.32) that for municipal and industrial wastewaters from the same area. Study of oil and grease discharges in an urban area around Jamaica Bay, New York (about one-half million persons), provided a ratio of 0.33 (Engineering Science, Inc., 1971). These results are consistent with those of a recent study in Stockholm (Söderlund and Lehtinen, 1970).

It is probable that the PHC fraction of oil and grease in runoff from urban areas (particularly those with heavy automobile traffic) is significantly higher than that in municipal and associated industrial wastewaters. We assumed that PHC represents 75 percent of the oil and grease in urban runoff, compared with 50 percent for oil and grease in municipal and industrial wastewaters. Thus, the expected ratios of PHC in urban runoff to that in wastewaters would range from 0.15 to 0.9 [(0.75) $\div$ (0.50) $\times$ the 0.1–0.6 range)]; this is best estimated at (0.75) $\div$ (0.50) $\times$ 0.33 or 0.5.

When these ratios are applied to the estimates of world PHC input from municipal and industrial (excluding refineries) wastewaters of 0.6 mta, the best estimate of world input of PHC from urban storm drainage is 0.3 mta with a probable range of 0.1–0.5 mta.

RIVER RUNOFF

The SCEP (1970) report estimated that the total United States input of oil in river runoff was about 0.15 mta. This calculation was based on an estimated river flow of 1.75 $\times$ 10^{12} tons per year and a mean oil concentration of 0.085 mg/liter. Review of the basis for the assumed mean oil concentration reveals a number of points for concern.

The concentration estimated was the mean of results [reported by the Federal Water Quality Administration (FWQA)] of carbon–chloroform extract (CCE) tests on 24 rivers. The standard CCE test as used by FWQA involved the filtration of subsurface water through a sand filter (to remove particulates) and an activated carbon absorption column. At the end of the filtration period the organic material adsorbed on the carbon is extracted with chloroform. Thus, the CCE test is a measure of all dissolved organic substances passing through the sand filter, retained on the activated carbon, and eluted with chloroform. No attempt was made to estimate the proportion of petroleum hydrocarbons in the materials measured by the CCE test.

A more serious criticism of the use of CCE data, however, is directed at the bulk of the potential PHC carried in river runoff being adsorbed on the silt load and/or carried on the water surface. In addition, the PHCs passing the sand filter would also be expected to pass through the carbon column if they properly represent the mineral oil fraction as claimed for the method of Lindgren (Lindgren, 1957).

The Mississippi River carries an annual silt load of about 5×10^8 tons (SCEP, 1970). Petroleum hydrocarbons in the sediments at the mouth of the Mississippi River have been measured by hexane extraction as about 400 mg/kg of sediment (dry basis) (McAuliffe, 1973b). The concentration level of hexane extractible material (HEM) of 400 mg/liter in the surface sediments at the mouth of the Mississippi River is supported by the data reported by Storrs et al. (1966) on surface sediments in San Francisco Bay and indicated a bay-wide mean of about 460 mg/kg. San Francisco Bay receives a large flux of sediment from the Sacramento and San Joaquin rivers, as well as considerable local PHC input. If this concentration is assumed to be that on the suspended silt in the river, then 0.2 mta of PHC will be carried out of the Mississippi River. On an overall basis (ignoring any PHC carried on the water surface), 0.2 mta in an average annual flow of 6×10^{11} m³ corresponds to a mean PHC concentration of 0.3 mg/liter. A bulk flow mean PHC concentration of 0.3 mg/liter appears to be a reasonable, if not low estimate, since higher concentrations have been reported for several rivers in the United States and Europe. For example, the dry weather flow oil concentrations in the Ventura and Santa Clara rivers in California have reported concentrations of about 4.0 mg/liter (Southern California Coastal Water Research Project, 1973); the bulk volume oil products concentration in the Danube River near Budapest has been reported in excess of 1 mg/liter (North-Transdanubia District Water Authority, 1973); and the median hydrocarbon concentration in the Rhine River near Koblenz is about 0.3 mg/liter (Hellman and Bruns, 1970). Thus, it appears that previous estimates of PHC input to the ocean from rivers may be too low.

The CCE concentration for the Mississippi River corresponds to the average CCE value for the total United States—0.085 mg/liter. Therefore, if the Mississippi River is considered representative of all United States rivers, its estimated PHC load can be multiplied by the ratio of total river flow to that of the Mississippi River, i.e., $1.6 \times 10^{12} – 6 \times 10^{11}$ mta or 2.7. This results in an estimated 0.53 mta of PHC carried to the ocean in United States rivers. If this figure is multiplied by 3 (the ratio of world to United States petroleum utilization), the world input would be about 1.6 mta.

FUTURE TRENDS

SEEPS

There is no evidence to indicate that assumed rates of natural seepage will change significantly over the time period considered herein.

OFFSHORE PRODUCTION

Weeks (1965) forecasts that by 1978 total worldwide offshore oil production would increase to 8,395 million barrels per year, or about 2.5 times that of 1971 production. Therefore, it is possible that the total hydrocarbon input from offshore producing operations could increase by a similar amount by 1978 and may reach 0.2 mta. In fact, a recent technology assessment of outer continental shelf oil and gas operations by the University of Oklahoma (Kash et al., 1973) suggests that no decrease in the percentage of composite oil and gas well blowouts offshore was obtained from 1964 through 1971. However, increased environmental concern, increased care, and improved equipment in all phases of offshore operations would reduce spillage by some degree below this projected level. This assumes that foreign offshore drilling and production practices remain comparable with those in the United States, a good assumption considering the requisite skills, capital investment, and severe financial loss involved when a drilling or production platform is lost offshore.

TRANSPORTATION LOSSES

Losses from Tankers during Normal Operations

The 1980 situation depends on decisions reached at the Intergovernmental Maritime Consultative Organization (IMCO) during their 1973 Marine Pollution Conference of 78 member nations. The IMCO conference developed a comprehensive study, *Convention for the Prevention of Pollution from Ships,* with provisions for annexes dealing with various types of pollutants, including oil and other noxious substances (Benkert and Williams, 1974). IMCO's mandatory annexes will be enforced in approximately 12 months, providing 15 states (representing at least 50 percent of the gross tonnage of the world's merchant shipping) have become parties to the convention. In this case, modern ships will be practicing LOT with reasonable efficiency and will be equipped for segregation of ballast; thus, resulting marine oil pollution should be negligible, that is, an approximate 95 percent reduction should result during normal operations. The convention also demands that oil loading terminals and repair facilities have reception facilities for tanker washings and dirty ballast. The total PHC contribution to the sea from tankers would, providing that the new international convention is strictly enforced, then be reduced from about 1.08 to 0.20 mta.

Terminal Operations

The quantity of oil handled will increase markedly to meet the western world's increasing demand for crude petroleum from overseas. Most of the spillage in terminals is caused by human error, and some improvement in this performance can be anticipated as the result of increased surveillance and cleanup penalties. The technology and capability for cleaning up small and medium-sized oil spills will also improve, such that there should be no major change in the total amount of spillage taking place during terminal operations; we estimate about 0.003 mta.

Losses from All Ships—Bilges, Bunkers, etc.

The volume of shipping will have increased, but reception facilities at ports for bilges will probably be more generally available, as will onboard separators. On balance, there should be some improvement, which we estimated as from 0.5 to 0.3 mta.

Tanker Accidents

We expect that designs to minimize spillage following accidents will be introduced in the new ships, and the capability to perform high seas cleanup enhanced. Also, navigational and communication aids will be improved, particularly with regard to the coastal confluence zones in the more heavily populated parts of the world. Marine vessel traffic systems will become much more common as an aid to transiting areas with unusually heavy ship traffic. There should be a slow downward trend in the average rate of loss from tanker accidents to about 0.15 mta, providing there is no catastrophic total loss of one or more very large crude carriers.

Nontanker Accidents

Although there should be improved navigational facilities, human error will remain the major factor, and little change is expected.

COASTAL REFINERIES

The trends toward recycling water coolant and the use of air cooling will continue, and legislative pressures will result in much more efficient oil removal from effluents. A steep reduction in the quantity of oil reaching the sea from this source, therefore, is to be expected, probably to a level of the order of 0.02 mta.

ATMOSPHERE

All data regarding atmospheric emissions of PHC indicate vehicle emissions are the single most important source. This influx may be considerably reduced by current measures to control automobile exhausts, ultimately reducing the input to the oceans. The U.S. National Air Pollution Control Administration has estimated that 1968 legislation standards will result in a decrease in vehicle hydrocarbon emissions to approximately one-third the 1968 values by 1985. Similar legislation on a worldwide basis, for vehicles as well as other sources, is required if the atmospheric input of PHCs to the oceans is to be controlled.

MUNICIPAL AND INDUSTRIAL (NONREFINERY)

It is probable that a substantial portion of the PHC found in municipal wastewaters comes from the dumping of waste oils. This practice is controllable to some degree, and with increasingly stringent environmental control regulations, there is some prospect for a reduction in PHC discharges from this source. The extent of this potential reduction, however, cannot be estimated until data are available on the magnitude of such inputs.

Similarly, most cities and regulatory agencies in the United States are becoming increasingly conscious of PHC inputs from industrial sources, and source control regulations will undoubtedly reduce the discharge of PHC to sewerage systems and receiving waters. The extent of such expected reduction is difficult to estimate, but a reduction in the next 5–7 years of 50 percent would seem reasonable. This would reduce the estimated industrial discharge (excluding refineries) in coastal areas from 0.3 to 0.15 mta. Such regulations would also result in some reduction in the estimated PHC inputs from river runoff, although it is not possible to estimate the amount.

URBAN STORM RUNOFF

Much of the PHC in urban storm drainage results from oil spillage on streets and highways and from fallout.

TABLE 1-7 Estimated Inputs of Petroleum Hydrocarbons in the Ocean during the Early 1980s[a]

Input Source	High	Modest	Low
Natural seeps		0.6	
Offshore production	0.2		
Transportation	0.8		
Coastal refineries	0.02		
Atmosphere			0.6
Municipal and industrial	0.45		
Urban runoff		0.3	
River runoff		1.6	
TOTAL	1.47	2.5	0.6
GRAND TOTAL		4.57	

[a] Input values are directly subject to global output values that may experience major shifts because of political, financial, economic, or exploration/production considerations.

Again, meaningful quantitative estimates of expected reductions resulting from increasing environmental concern and regulation cannot be made until source generation rates are quantified.

RIVER RUNOFF

Reductions in inputs from river flows depend on reducing the amount of PHC entering natural streams in wastewaters and from the atmosphere. Assessment of potential reductions in river inputs is further compounded by the many complex processes affecting PHC in river systems.

CONCLUSIONS

In Table 1-7 various estimated inputs presented above have been categorized according to the degree of confidence or uncertainty in our estimates. Although all of these estimates have been derived from a small data base and partially supported assumptions, almost one-half the total estimated inputs appears to be reasonably realistic and supportable. We have identified a number of major uncertainties and data needs, which are discussed briefly in the following sections.

ATMOSPHERIC DATA

The principal factors affecting the estimates of PHC inputs from the atmosphere include the following:

1. Reaction kinetics of the various hydrocarbon compounds entering the atmosphere;
2. Nature and fates of the various reaction products; and
3. Nature and distribution of particulate materials, their absolute and relative distributions in relation to major areas of hydrocarbon generation, and relationships among numbers, size distributions, and mass fallout rates of particulates.

There is a dearth of information concerning these factors and additional data are required before more accurate, quantitative estimates can be prepared. Background papers by Feuerstein (1973), Hallhagen (1973), McAuliffe (1973a), and Duce (1973) prepared on this subject cite the difficulty of quantitatively defining the total PHC atmospheric rainout cycle, particularly in the presence of biogenically related hydrocarbons, which are continually being formed by both terrestrial and marine plants in amounts that appear to be two or three orders of magnitude greater than the PHC atmospheric rainout values.

RIVER RUNOFF

The estimate for river runoff input is directly dependent on the estimated concentration of PHC on the suspended particulate matter in the river flow. For this report the assumed concentration was based on measured PHC concentrations in sediments deposited at the mouth of the Mississippi River. Direct measurements of PHC on suspended particulates should be made. Additional factors, not included in our estimates but which should be measured, include PHC dissolved and dispersed in river waters and that carried on the water surface.

NATURAL SEEPS

The major uncertainties connected with estimates of worldwide inputs from natural seeps are completeness of the inventory of seep areas and measurement of seepage rates. Such techniques as infrared and ultraviolet surveillance have been developed that could aid in mapping seep occurrences. In addition, seismic detection of gas bubbles in the water column may also be indicative of oil seepage in some areas. To date, no satisfactory methods are available for measuring rates of flow from natural seeps; however, much better estimates of inputs from seeps could be made if such methods were developed.

SAMPLING AND ANALYSIS

One of the greatest sources for error in the estimates given herein is inadequate sampling and analysis procedures. It is necessary to develop suitable techniques to sample water, wastewater, air, and sediments in such a manner that those samples are representative of existing conditions. Similarly, analytical procedures must be developed to measure PHC content in each of the sources sampled. Perhaps the most acute problem is associated with surface films. It is necessary to assess quantitatively the amount of film, as well as its associated PHC content.

RECOMMENDATIONS

An obvious question remains, "What can be done to realistically reduce the inputs of PHC to the oceans?"

We recommended additional effort in three areas deemed appropriate:

• Improve operation and control of tanker and shipping operations with special emphasis on achieving maximum LOT operation.
• Improve control of municipal and industrial wastewater treatment processes, and improve control and regulation of PHC sources in sewerage systems.
• Reduce PHC emissions from automobiles and other internal combustion engines.

Appendix: Characterization of U.S. Statistics for Petroleum Accidentally Released Into the Sea

Detailed global statistics for the amount of petroleum hydrocarbons released into the ocean either directly or by secondary routes do not yet exist. The world data base for such releases is patchy and not standardized among countries, many of which have no reporting method whatsoever. Regions subject to oil outflows due to tanker accidents in 1969 and 1970 have been briefly analyzed by Porricelli and Keith (1973).

TABLE A-1 Geographic Distribution of Polluting Spills into U.S. Waters During 1971[a]

Geographic Locale	Number of Incidents	Percent of Total	Weight of Spills[b] (metric tons)	Percent of Total
Inland waters				
Roadsteads	99	1.1	47	0.2
Ports	63	0.7	581	2.0
Terminals and docks	141	1.6	255	0.9
Beaches	7	0.1	121	0.5
River areas	252	2.9	2,281	8.0
Nonnavigable areas	69	0.8	1,270	4.4
TOTAL INLAND WATERS	631	7.2	4,555	16.0
Coastal waters (including the Great Lakes)				
Bays, estuaries, and sounds	2,933	33.6	6,835	23.9
Ports	1,452	16.6	2,621	9.1
Terminals and docks	869	10.0	1,530	5.3
Beaches	93	1.1	56	0.2
Channels, canals, and inlets	869	10.0	1,389	5.6
River areas	938	10.7	2,075	7.3
Nonnavigable areas	47	0.5	7,060	24.6
Open waters (Great Lakes or territorial sea)	315	3.6	122	0.4
TOTAL COASTAL WATERS	7,516	86.1	21,688	76.4
Contiguous zone	396	4.5	2,104	7.4
High seas	193	2.2	66	0.2
TOTAL	8,736	100.0	28,413	100.0

[a] From *Polluting Spills in U.S. Waters, 1971*, internal report of the U.S. Coast Guard, Washington, D.C. (date unknown).
[b] About 1 percent by weight of these spills is either unknown as to type of material, or is sewage, refuse, dredge spoil, or other material.

In addition, considerable information of this type is becoming available in the United States as a result of Section 311.b.5 of the Federal Water Pollution Control Act of 1970. This section requires that all accidental releases of petroleum be reported to the appropriate agency of the U.S. Government, which, in most cases, is the U.S. Coast Guard. These reported data are then compiled via a computerized Pollution Incident Reporting System into an annual published base. To date, these annual tabulations exist for part of 1970 and for all of 1971. Even today, however, despite fiscal penalties for nonreporting, there are many mystery spills, perhaps on the order of 30 percent, in which neither the spiller nor the amount is known. In the coming years, we anticipate that data will be complete because people will become aware of the requirements to report discharges of oil (in harmful quantities) to the U.S. Coast Guard. Thus, the summation of reported spills may be considered as giving a very conservative value for the amount of petroleum-related products spilled.

The geographic distribution of 1971 spills in the United States is shown in Table A-1. Of approximately 8,700 known spills, nearly two-thirds occurred in ports, bays, and sounds. The distribution by size of spills during 1970 is shown in Table A-2. Catastrophic spills, each over 3,250 metric tons, contributed nearly two-thirds of the volume spilled. The 3,648 reported spills of less than 32 metric tons each contributed only 5 percent of the total volume. Types of petroleum hydrocarbons spilled during 1971 are shown in Table A-3, where it will be noted that there were about equal amounts of light, heavy, and other types of oils spilled during the year.

TABLE A-2 Numerical Distribution of Water Polluting Spills of Petroleum Hydrocarbons in the United States During 1970 by Size[a]

Weight of Spill (metric tons)	Number of Spills	Total Weight of Petroleum Spilled (metric tons)	Percent of Total Weight Spilled
>3,250	4	32,200	65
325–3,250	14	10,400	21
32–325	45	4,400	9
<32	3,648	2,500	5
TOTAL	3,711	49,500	100

[a]From *Polluting Spills in U.S. Waters, 1971*, internal report of U.S. Coast Guard, Washington, D.C. (date unknown).

TABLE A-3 Types of Petroleum Hydrocarbons Spilled into U.S. Waters During 1971[a]

Type of Petroleum Hydrocarbons	Number of Incidents	Percent of Total	Weight of Spills (metric tons)	Percent of Total
Light oil gasoline, light fuel oil, kerosene, light crude	4,320	59	9,150	33
Heavy oil diesel oil, heating oil, heavy fuel oil, heavy crude, asphalt	1,603	21	9,500	34
Lubricating oil	168	2	72	0.3
Waste oil	930	12	535	2
Other oil	462	6	8,650	31
TOTAL	7,483	100	27,907	100.3

[a]From *Polluting Spills in U.S. Waters, 1971*, internal report of U.S. Coast Guard, Washington, D.C. (date unknown).

References

Altschuller, A. P. 1966. Reactivity of organic substances in atmospheric photo-oxidation reactions. Air Water Pollut. Int. J. 10:713.

American Public Health Association. 1971. Standard methods for the examination of water and wastewater. 13th ed. Washington, D.C.

Benkert, W. M., and D. H. Williams. 1974. The impact of the 1973 IMCO Convention on the Maritime Industry. Marine Technology 11(1):1–8.

Blokker, P. C. 1971. Prevention of water pollution from refineries. Proceedings, Seminar on Water Pollution by Oil, Aviemore, 1970. Institute of Petroleum, London.

Blumer, M. 1972. Submarine seeps: Are they a major source of open ocean oil pollution? Science 176:1257–1258.

British Petroleum Co., Ltd. 1971. BP statistical review of the world oil industry. London.

Brummage, K. G. 1973a. Sources of oil entering the sea. Background papers, workshop on petroleum in marine environment. Ocean Affairs Board, National Academy of Sciences, Washington, D.C.

Brummage, K. G. 1973b. What is marine pollution? Symposium on Marine Pollution. Royal Institute of Naval Architects, London.

Duce, R. A. 1973. Atmospheric hydrocarbons and their relation to marine pollution, pp. 416–430. Background papers, workshop on petroleum in marine environment. Ocean Affairs Board, National Academy of Sciences, Washington, D.C.

Engineering Science, Inc. 1971. Final Report-year 1, Spring Creek Auxiliary Water Pollution Control, Jamaica Bay Study. Bureau of Water Pollution Control, Department of Water Resources, City of New York.

Farrington, J. W., and J. G. Quinn. 1973. Petroleum hydrocarbons and fatty acids in wastewater effluents. J. Water Pollut. Control Fed. 45:704.

Feuerstein, D. L. 1973. Input of petroleum to the marine environment, pp. 31–38. Background papers, workshop on petroleum in marine environment. Ocean Affairs Board, National Academy of Sciences, Washington, D.C.

Hallhagen, A. 1973. Survey of present knowledge and discussion of input of petroleum into the marine environment in Sweden, pp. 39–40. Background papers, workshop on petroleum in marine environment. Ocean Affairs Board, National Academy of Sciences, Washington, D.C.

Hellman, von Hubert, and F. J. Bruns. 1970. Uter Suchungrn Zur Kohlenwasserstoffracht des Rheins 1968/69 and Uberlegungen Zu Deren Herkunft, Deutsche Gewasserkundl Mitteilung, pp. 14–18. No. 1.

Holdsworth, M. P. 1971. Oil pollution at sea. Symposium on Environmental Pollution. University of Lancaster, U.K.

Intergovernmental Maritime Consultative Organization (IMCO). 1972a. La pollution accidentelle des mers par hydrocarbutes. Tome II, Etude analytique des accidents. IMCO paper MPXIII/2(a)/6. Ecocentre Report, Paris.

IMCO. 1972b. Report on study IV. Clean tanks for ballast prior to vessel sailing. IMCO paper MPXIII/2(a)/6.

IMCO. 1973. Environmental and financial consequences of oil pollution from ships. Report on study VI. Preparations for international marine pollution conference, 1973, IMCO, London. Appendix 1.

Junge, C. E. 1963. Air Chemistry and Radioactivity. Academic Press, New York.

Kash, D. E., et al. 1973. Energy Under the Oceans. A Technology Assessment of Outer Continental Shelf Oil and Gas Operations. University of Oklahoma Press, Tulsa. 378 pp.

Lindgren, C. G. 1957. Measurement of small quantities of hydrocarbons in water. J. Am. Water Works Assoc. 49:55–62.

McAuliffe, C. D. 1973a. Partitioning of hydrocarbons between the atmosphere and natural waters, pp. 380–390. Background papers, workshop on petroleum in marine environment. Ocean Affairs Board, National Academy of Sciences, Washington, D.C.

McAuliffe, C. D. 1973b. Chevron Oil Field Research Co. (personal communication).

McCaslin, J. 1972. Worldwide offshore oil output nears 9 million B/D. Oil Gas J. 70(17):196–198.

National Air Pollution Control Administration. 1968. Nationwide inventory of air pollutant emissions, 1968. Raleigh, N.C.

Newell, R. E. 1971. The global circulation of atmospheric pollutants. Sci. Am. 224(1):32–42.

North-Transdanubia District Water Authority. 1973. Project Hungary 3101 pilot zones of water quality management. 7 pp.

Porricelli, J. D., V. F. Keith, and R. L. Storch. 1971. Tankers and the ecology.

Porricelli, J. D., and V. F. Keith. 1973. An analysis of oil outflows due to tanker accidents, pp. 3–14. Proceedings, Joint Conference on the Prevention and Control of Oil Spills. American Petroleum Institute, Washington, D.C.

Robinson, E., and R. C. Robbins. 1968. Sources, abundance and fate of gaseous atmospheric pollutants. American Petroleum Institute, Washington, D.C.

Söderlund, G., and H. Lehtinen. 1970. Is dumping of snow into lakes a pollution problem? Vatten 26(2):146–148.

Southern California Coastal Water Research Project. 1973. The ecology of the Southern California bight; implications for water quality management. Three year report. Los Angeles, Calif. 531 pp.

Speers, G. C., and E. V. Whitehead. 1969. Organic Geochemistry. Edited by G. Eglinton and M. T. G. Murphy. Springer, New York. Chapter 27.

Storrs, P. N., E. A. Pearson, and R. E. Selleck. 1966. A comprehensive study of San Francisco Bay. Final report. Vol. V, U.C. Sanitary Engineering Research Laboratory Publication No. 67-2. 140 pp.

Storrs, P. N. 1973. Petroleum inputs to the marine environment from land sources. Background papers, workshop on petroleum in marine environment. Ocean Affairs Board, National Academy of Sciences, Washington, D.C.

Study of Critical Environmental Probe (SCEP). 1970. Man's impact on the global environment. Assessment and recommendations for action. MIT Press, Cambridge, Mass.

U.S. Coast Guard. 1973. Draft environmental impact statement. For International Convention for the Prevention of Pollution from Ships, 1973. 89 pp., 6 appendices.

U.S. Department of the Interior. 1972a. Draft environmental statement of the proposed 1972 outer continental shelf oil and gas general lease sale. Offshore Louisiana. U.S. Department of the Interior, Washington, D.C.

U.S. Department of the Interior. 1972b. Draft environmental statement of the proposed 1973 outer continental shelf oil and gas general sale. East Texas. Section III, No. E, pp. 246–253.

U.S. Department of the Interior. 1973. Estimated international flow of petroleum and tanker utilization, 1971–72. Office of Oil and Gas, U.S. Department of the Interior, Washington, D.C.

Victory, G. 1973. The load on top system, present and future, pp. 10–20. Symposium on Marine Pollution. Royal Institute of Naval Architects, London.

Wilson, R. D., P. H. Monaghan, A. Osanik, L. C. Price, and M. A. Rogers. 1973. Estimate of annual input of petroleum to the marine environment from natural marine seepage. Background papers, workshop on petroleum in marine environment. Ocean Affairs Board, National Academy of Sciences, Washington, D.C.

Weeks, L. C. 1965. World offshore petroleum resources. Am. Assoc. Pet. Geol. Bull. 49:1680–1693.

2 Analytical Methods

Chemical

No one analytical technique can solve all types of problems concerned with the determination of hydrocarbons in the marine environment. As will be apparent from this chapter, as well as from techniques cited throughout this publication, diagnostic methods are required to answer many questions. Hydrocarbon concentrations range from pure oil to the extremely low levels of a combination of hydrocarbons or specific hydrocarbons in the open ocean. Methods must differentiate hydrocarbons into classes, such as alkanes or aromatics; molecular weight categories; specific hydrocarbons or known polycyclic aromatic hydrocarbons (such as 3,4-benzopyrene). Techniques must be available for determining these hydrocarbons in the atmosphere, waters, sediments, and biological material. Suitable techniques must be available to determine the fate of hydrocarbons discharged to the marine environment from whatever sources (spills, chronic discharges, municipal discharges, or biogenesis).

Biological studies of the effects of hydrocarbons on marine organisms must be accompanied by chemical analyses that support the conclusions drawn. This entails at least the analysis of the organisms and their natural or laboratory habitat (water, sediment, etc.) for concentration and composition of hydrocarbons.

General Considerations

It is essential to realize that crude oils can contain many tens of thousands of compounds. Also, crude oils from different regions can vary greatly (Smith, 1968).

Crude oils consist principally of hydrocarbons and their derivatives, with appreciable amounts of combined sulfur, some nitrogen, and less oxygen. In addition, very small concentrations of metals, including nickel, vanadium, and iron, are complexed with natural organic chelates such as porphyrins. The percentages of sulfur and nitrogen in the higher boiling fractions of some crude oils are often so high that the majority of the larger molecules probably contain some heteroatoms.

HYDROCARBON TYPES IN CRUDE OILS

All crude oils contain three general classes of hydrocarbons—alkane, cycloalkane, and aromatic—but not alkene (olefin) hydrocarbons.

Normal alkanes from methane to beyond C_{60} have a ratio of abundance of about one for odd- to even-numbered carbon chains. Many parallel, homologous branched-chain alkanes are present, including a series of isoprenoid alkanes.

A complex mixture of five- or six-carbon, hydrogen-saturated ring structures are present in crude oil with the alkyl-substituted hydrocarbons being more abundant than their parent hydrocarbon.

Aromatics include six-carbon unsaturated ring structures such as benzene, and a complex mixture of mono-, di-, tri-, and tetraalkyl benzenes, naphthalenes, and polynuclear aromatic hydrocarbons with multiple alkyl substitutions. Also included in this class are hydrocarbons sometimes designated as naphthenoaromatics, because of the mixed nature of the molecule—i.e., part aromatic, part cycloalkane.

HYDROCARBON TYPES IN REFINED PETROLEUM PRODUCTS

Various refined petroleum products, such as gasoline, jet fuel, kerosene, diesel, distillate fuels, and residual fuels contain all of the classes of hydrocarbons found in crude oils. Alkenes (olefins) may occur at concentrations up to 30 percent in gasoline from cracking stocks and at about 1 percent in jet fuel. Otherwise, they are not found in refined products other than in trace amounts.

BIOGENIC HYDROCARBONS

Terrestrial and marine organisms contain hydrocarbons. These organisms either make their own hydrocarbons, obtain them from their food source, or convert precursor compounds obtained with their food. An example of the latter is the conversion of phytol to pristane (Blumer, 1967). A comprehensive review of hydrocarbons native to organisms was made by Gerarde and Gerarde (1962). Clark (1966) reviewed the literature on saturated hydrocarbons.

Both terrestrial and marine organisms synthesize normal alkanes, predominately with odd-numbered carbon chains. In many instances one or two odd-carbon numbered normal alkanes predominate. In marine phytoplankton normal alkanes with 15, 17, 19, and 21 carbon atoms are most abundant (Clark, 1966; Blumer et al., 1971). In marsh grasses and Sargassum, C_{21}–C_{29} n-alkanes predominate. In terrestrial materials such as plant waxes (grass, leaf, and fruit waxes), the normal alkanes with 27, 29, 31, and 33 carbon atoms are most abundant (Clark, 1966). Some bacteria have been found to contain equal amounts of even- and odd-carbon-numbered C_{25}–C_{32} n-alkanes (Davis, 1968). Marine sponges and corals have also been shown to have little if any odd-carbon-number preference for C_{25}–C_{34} n-alkanes (Koons et al., 1965).

Branched alkanes, including pristane, have been found in organisms. In some fishes pristane is the most abundant (Blumer, 1967). Several monomethyl branched alkanes (Youngblood et al., 1971) have been found in organisms and an alkylcyclopropane has been tentatively identified (Youngblood et al., 1971; Han and Calvin, 1969).

Alkenes are often a major portion of the hydrocarbons found in organisms, particularly marine organisms. Squalene is the major hydrocarbon constituent of basking shark liver, shark liver oil, and cod liver oil. Isoprenoid C_{19} and C_{20} mono-, di-, and triolefins are present in copepods and some fish (Blumer et al., 1969). Several straight-chain mono- to hexaolefins have been found in considerable quantities in many organisms (Blumer et al., 1970; Youngblood et al., 1971, 1973).

Cycloalkanes and Cycloalkenes, hydrocarbons containing one to three nonaromatic rings, are present in a number of terrestrial plants (Gerarde and Gerarde, 1961, 1962). Most are classified as terpenes because of their structural relationships to isoprenes. Terpenes are released to the atmosphere in large amounts from conifers (Rasmussen and Went, 1965).

Aromatic hydrocarbons have been isolated from terrestrial plants and spices (Gerarde and Gerarde, 1962). ZoBell's (1971) review of polynuclear aromatic hydrocarbons in the marine environment suggests that some are synthesized by marine microorganisms. The ubiquity of some polynuclear aromatic hydrocarbons in the environment suggest indigenous formation in plants and microorganisms, particularly in a wide variety of materials not likely to have been associated with pyrolitic processes (Andelman and Suess, 1970). All the aromatic hydrocarbons are in extremely low concentrations, generally less than 1 percent of the total hydrocarbons of marine organisms analyzed to date.

Only a limited number of terrestrial and marine species, from only a few geographical locations, have been analyzed for their native hydrocarbons. Also, many investigators have limited their analytical techniques to searching only one or two classes of hydrocarbons, usually alkanes and alkenes, and would not have detected other classes if present. It may be that other classes of hydrocarbons are more prevalent in nature than the limited analyses suggest.

DIFFERENTIATION OF PETROLEUM HYDROCARBONS FROM BIOGENIC HYDROCARBONS

Petroleum and biogenic hydrocarbons can be distinguished as follows, thus providing useful means for detecting petroleum. Note that not all differences apply to all organisms, nor to all crude oils and refined products.

1. Petroleum contains a much more complex mixture of hydrocarbons, with much greater ranges of molecular structure and weight.

2. Petroleum contains several homologous series, with adjacent members usually present in nearly the same concentration. The approximate unity ratio for

even- and odd-numbered alkanes is an example, as are the homologous series of C_{12}–C_{22} isoprenoid alkanes. As previously mentioned, marine organisms have a strong predominance of odd-numbered C_{15} through C_{21} alkanes.

3. Petroleum contains more kinds of cycloalkanes and aromatic hydrocarbons. Also, the numerous alkyl-substituted ring compounds have not been reported in organisms. Examples are the series of mono-, di-, tri-, and tetramethyl benzenes and the mono-, di-, tri-, and tetramethyl naphthalenes.

4. Petroleum contains numerous naphthenoaromatic hydrocarbons that have not been reported in organisms. Petroleums also contain numerous heterocompounds containing S, N, and O, metals, and the heavy asphaltic compounds.

Certain analytical techniques are superior to others in differentiating hydrocarbons from petroleum and biogenic sources. For example, gas chromatography readily shows the differences in *n*-alkane distributions typical of biogenic and petroleum derived hydrocarbons. In contrast, infrared analysis is of little value in quantifying these differences.

Analysis of Oils

At the time of an oil spill, samples should be taken from the source oils, water surface and collected if stranded on beaches. Oil on water or beach samples should be collected with time to follow weathering sequences. These samples of oil serve as reference material for subsequent correlation of hydrocarbons that may be detected in waters, sediments, and tissue samples.

Many analytical techniques are cited in this section but only those that are used for water, sediment, and tissue analysis are normally undertaken. Although many of the analytical techniques cited have application for surveillance, this is outside the scope of the workshop, and specific techniques as applied for surveillance have not been included.

Sample Collection

A relatively large sample of each oil makes possible more accurate analysis and is likely to be more representative than a small sample. A quart sample of the original petroleum should be collected, along with similar amounts of oil from the water's surface and, if the oil reaches, from beaches and shores. Water collected with the oil should be kept at a minimum, to reduce bacterial activity and, if the samples are to be frozen, breakage. If bioassays are to be undertaken, larger amounts (several gallons) of oil must be collected.

Kawahara (1969) described the use of commonly available devices for collecting oil. Kreider (1971) recommends skimming oil into a container, when the amount of spilled oil is appreciable, or collecting thin oil films using treated glass cloth. At the time of a major oil spill, skimming devices used to recover oil can provide samples.

At all times care must be taken to avoid contamination with other oily materials or oil-soluble substances that are not separable from oil by the analytical procedures.

The exact collection location, conditions, date, time, and name of collector should be recorded. Sample identification tags must be durable and oilproof.

Sample Preservation

Preservation of the oils from a spill involves containment of low boiling components and protection against oxidation and microbial attack. Crude oils and refined products containing less than 3 percent water directly from tankers or pipelines should be preserved by sealing in glass bottles and storing at ambient temperature, upright and in the dark (Kawahara, 1969).

Crude oils from slicks and producing wells may contain appreciable water. If possible, the bulk of the water should be removed and the sample container frozen and maintained at -20 °C until the time of analysis to minimize bacterial alteration of the oil. An alternative procedure is to add about 1 part in 10,000 of mercuric chloride to the water phase to inhibit bacterial activity. The sample can then be stored at ambient temperatures. This alternative procedure may be preferred for oils that may become heterogeneous when held at -20 °C. Microbiological alteration of oils is discussed in detail in a later chapter. ZoBell (1969) indicates that more than 100 species of bacteria, yeast, and fungi have been found to oxidize one or more hydrocarbons; different classes of hydrocarbons are differentially attacked by microorganisms.

ASTM subcommittee D-19.10 has prepared a tentative method for preserving samples that is based on the experience of many workers.

Preparation of Oil Samples for Analysis

Oil samples containing little water and no extraneous material can be analyzed directly. However, those collected from the sea surface or from beaches usually contain appreciable water and extraneous solids, from which the oil must be isolated by solvent extraction or by distillation. Kreider (1971) recommends dissolving the sample in chloroform, centrifuging, and removing the chloroform by evaporation. However, many of the volatile hydrocarbons are lost by this procedure.

ANALYTICAL METHODS

The principal techniques used to analyze oils are summarized in Table 2-1 and in the subsequent section for measurement of hydrocarbons in water. The techniques in Table 2-1, to be discussed briefly, are generally suitable only for separate-phase oil.

Routine Tests

If sufficient sample is available, bulk properties such as specific gravity, viscosity, wax, and asphaltene contents can be determined. These and other methods have been reviewed by Duckworth (1971). He states that a high asphaltene content may be useful for distinguishing residual fuel oils from weathered crude oils.

Gel Permeation Chromatography

Separations by gel permeation chromatography are based mainly on molecular size. The technique has become widely used to define physical parameters such as molecular weights. Done and Reid (1970) describe a relatively simple and short technique for distinguishing crude oils. Albaugh and Talarico (1972) describe a gel permeation technique that uses three detectors to characterize crude and fuel oils.

Nuclear Magnetic Resonance Spectroscopy

Proton nuclear magnetic resonance (NMR) is not likely to be of value as a discriminating tool because the signal from the hydrogen atom on aromatic and unsaturated carbon atoms and the hydrogen on saturated carbon atoms tend to lump together. However, ^{13}C NMR offers more promise because greater discrimination of types between these two signals is available. A review (Clutter et al., 1972) on the use of this technique indicates that this technique gives the information on the degree of substitution of aromatic structures, the complexity of aliphatic structures, and the amounts of heteroatoms.

Carbon and Sulfur Isotope Ratios

A multiparameter approach to oil identification (Miller, 1973) has been used to distinguish various oils by their isotopic composition of ^{13}C/^{12}C and ^{34}S/^{32}S. Koons et al. (1971) reported on the use of ^{13}C/^{12}C isotope ratio, gas chromatography, and mass spectroscopy for identifying oils. Isotope ratio measurements have considerable diagnostic power, but require specialized and dedicated equipment (McKinney et al., 1950; Thode et al., 1958; Hoefs, 1973).

Kjeldahl, Microcoulometer

Nitrogen content can be of assistance in distinguishing various oils, particularly those having larger amounts of high-molecular-weight heteroatoms. Kreider (1971) reported obtaining data on several samples of oil by the Kjeldahl procedure. Nitrogen can be determined with good precision and is useful with residual fuel oils and certain crude oils. The microcoulometric procedure (Martin, 1966) for total nitrogen is more rapid, and a sample smaller than is required by the Kjeldahl method is sufficient.

X-Ray Fluorescence, Atomic Absorption, and Fluorescence Spectroscopy

Sulfur, nickel, vanadium, iron, and other elements can be determined by these techniques. Sulfur content has often been used to categorize various crude and residual fuel oils. Nickel and vanadium contents, long recognized as characteristic properties, can also be determined by classical chemical analysis. These methods are inexpensive in terms of capital equipment, but expensive in terms of operating costs. The reverse is true of instrumental methods such as X-ray fluorescence.

Emission Spectroscopy and Neutron Activation Analysis

These techniques analyze many trace metals. Neutron activation analysis (Bryan et al., 1970; Lukens et al., 1971) is more sensitive than emission spectroscopy, but in the less than per million concentration range one cannot be sure that the indication is not due to contamination. This is particularly true where the sample has been in contact with seawater or sediment. Also, multielement neutron activation analysis is less readily available than emission spectroscopy.

HYDROCARBONS IN WATER

As discussed previously, hydrocarbons from crude oil and refined products are diverse mixtures of classes whose individual compounds cover a wide range of volatility. Crude oils, for example, may contain hydrocarbons over a carbon-number range from 1 to over 70. A given analytical method will not usually be able to

TABLE 2-1 Analytical Techniques Suitable for Separate Phase Oil Only[a]

Technique	Component Determined	Sample Size (g)	Advantages	Disadvantages	Equipment (Approximate Cost in Dollars)	Analysis Time: Operator (Elapsed)	References
Routine tests	Specific gravity, viscosity, wax, asphaltene of whole oil	1–10	Rapid, simple, routine tests	Only useful if samples are unweathered	Laboratory (500)	10–60 min (10–60 min)	Duckworth, 1971
Gel permeation chromatography	Molecular weight separation, C_{10}–C_{max}	10^{-2}–10^{-3}	Can be used as a preparative method for other analytical methods.	Limited diagnostic ability	Liquid chromatograph (12,500)	10 min (60 min)	Albaugh and Talarico, 1972; Done and Reed, 1970
Nuclear magnetic resonance spectrometry (^{13}C)	Alkane, aromatic and heteroatoms, C_{10}–C_{max}	10^{-1}	Rapid. Measure degree of aromatic substitution.	Expensive. Limited diagnostic ability	^{13}C NMR spectrometer (80,000)	15 min (60 min)	Clutter et al., 1972
Carbon and sulfur isotope ratios	$^{13}C/^{12}C$, $^{34}S/^{32}S$, C_6–C_{max}	10^{-3}	Considerable diagnostic power	Specialized, dedicated equipment not readily available	Isotope ratio mass spectrometer (75,000)	1 h (1 h)	McKinney et al., 1950; Thode et al., 1958; Hoefs, 1973
Kjeldahl, microcoulometer	Nitrogen, C_6–C_{max}	0.2	Standard classical method, rapid with automatic readout	Slow, considerable operator skill	Laboratory glassware, furnace, and microcoulometer (500–9,000)	2 h 10 min (10 min)	Kreider, 1971; Martin, 1966
X-ray fluorescence	S, Ni, V, Fe, and other elements if required, C_{11}–C_{max}	1–5	Rapid, multielement	Matrix effects, limited sensitivity	X-ray spectrometer (25,000)	10 min (10 min)	
Atomic absorption and fluorescence spectroscopy	Ni, V, and other elements if required, C_{11}–C_{max}	1–5	High sensitivity	Ashing may be necessary, one element at a time, slow	Respective instruments (5,000–10,000)	2 h (2 h) per element	
Emission spectroscopy	Many trace metals, C_{11}–C_{max}	1	High sensitivity, simultaneous determination of many elements	Ashing necessary, sensitivity lower than neutron activation	Emission spectrometer (25,000–75,000)	2 h (2 h)	
Neutron activation	Many trace elements, C_{11}–C_{max}	0.1	Very high sensitivity, simultaneous determination of many elements	Equipment cost very high—usually a service from central organization		30 min (1 wk or more for sample irradiation for some elements)	Bryan et al., 1970; Lukens et al., 1971

[a] See Table 2-2 for additional methods.

measure hydrocarbons over the entire volatility range. In addition, special methods may be required to measure the very low concentrations and the wide spectrum of hydrocarbons possible in open ocean waters, for example. An analytical method that is intended to measure volatile hydrocarbons may be different than one for measuring the higher molecular weight hydrocarbons. Thus, more than one procedure is required to measure the whole range of hydrocarbons in a given water sample.

SAMPLE COLLECTION AND PRESERVATION

Low-Molecular-Weight Hydrocarbons (C_1–C_{10})

Care must be taken during collection and preservation to avoid loss of the low-molecular-weight hydrocarbons, which can readily escape to the atmosphere. Water samples can be collected from all depths with hydrographic equipment and various sampling devices such as a Van Dorn water sampler. Rubber seals on the sampling devices apparently are satisfactory when the waters are to be analyzed for these hydrocarbons. The retrieved water should be transferred through clean Teflon tubing with a minimum of agitation into a clean glass bottle suitable for sealing by means of a cap with a Teflon liner. A preservative amount of mercuric chloride or sodium azide should be added before quickly capping the bottle. When capped properly, a gastight seal is obtained.

High-Molecular-Weight Hydrocarbons (C_{11} plus)

The determination of C_{11} plus hydrocarbons at the levels in ocean water is much more subject to contamination than for low-molecular-weight hydrocarbons. Therefore, extreme care must be taken to use proper sampling equipment and carefully cleaned collection and storage vessels. A particularly desirable sampling device is constructed only of glass, stainless steel, and Teflon.

The retrieved samples should be transferred through clean Teflon tubing into sample bottles that have been cleaned previously with carbon tetrachloride. The bottles should be from 0.5 to 20 liters, depending on the sensitivity required and an appropriate volume of hydrocarbon-free CCl_4 should be added to each sample. For certain analyses, it will also be appropriate to add hydrocarbon-free sodium chloride and hydrochloric acid.

The collection bottles should be capped immediately, using a clean Teflon liner, and the sample should be stored upright. Immediately after collection, the sample should be shaken in order to initially extract the hydrocarbons into the CCl_4.

ANALYTICAL METHODS FOR C_1–C_{10} HYDROCARBONS

Gas Equilibration

Low-molecular-weight hydrocarbons in waters from all sources can be measured using the gas equilibration technique of McAuliffe (1971). Alkane, alkene, and cycloalkane hydrocarbons can be determined to a sensitivity level of 1–3 parts per trillion (10^{12}) by weight; aromatic hydrocarbons, to 8–10 parts per trillion. Because the method separates hydrocarbons qualitatively from water soluble organics (alcohol, acids, aldehydes, etc.), hydrocarbons can be directly measured in the presence of such nonhydrocarbons without sample preparation.

Gas Stripping

The gas equilibration method will measure background levels of methane in open seawaters, but has insufficient sensitivity to do so for ethane, ethene, propane, and propene (less than 1 ppt). Swinnerton and Linnenbom (1967) have measured these hydrocarbons in open ocean waters by gas-stripping 1-liter samples of water. Other than for these very light gases, the gas-stripping technique can confuse hydrocarbons with nonhydrocarbons that may be present in the water.

Vacuum Degassing

Low-molecular-weight hydrocarbons through C_4 have been continuously monitored by marine seep detectors, that separate dissolved gases from seawater by vacuum. These systems have been described (Schink et al., 1971; Fort et al., 1973), and a system is offered commercially by InterOcean, San Diego. The systems have excellent sensitivity, because dissolved oxygen and nitrogen contribute about 50 ml gas/liter water. Because of the favorable partitioning of low-molecular-weight hydrocarbons into the gas phase, these detectors can measure open ocean background levels of methane, ethane, ethene, propane, propene-iso, and normal butane (in the subpart per trillion range for C_2–C_4 hydrocarbons). Analysis for these hydrocarbons from nearsurface down to 200 m occurs every 3 min as the ship moves at up to 10 knots. Although no preservation of the sample is necessary and analysis is very much less subject to hydrocarbon contamination, the systems are expensive.

The above discussion emphasizes techniques suitable for measuring the very low concentrations of low-molecular-weight hydrocarbons in water samples. Higher concentrations can be measured easily by gas equilibration. For example, the method can be used to determine the soluble low-molecular-weight hydro carbons in water that have been equilibrated against crude oils and refined products (McAuliffe, 1969, 1971). This technique, along with that of Boylan and Tripp (1971), can provide additional valuable information in distinguishing various crude oils and refined products. Water equilibrated against various oils and refined products permits separation of the aromatic hydrocarbons from all other hydrocarbons, and composition of these aromatics can be helpful in distinguishing various oils.

ANALYTICAL METHODS FOR C_{11} PLUS HYDROCARBONS

As with low-molecular-weight hydrocarbons, techniques for measuring high-molecular-weight hydrocarbons require sufficient flexibility to measure rather high concentrations in water to the very low concentrations observed in the open ocean. The procedures must also separate hydrocarbons from solvent extractable organic nonhydrocarbons. Seawater contains about 1 ppm soluble organic compounds, expressed as carbon. This material is from living and dead marine organisms, part of which CCl_4 extracts with the hydrocarbons present. The efficiency of CCl_4 extraction of hydrocarbons from seawater has been shown by Simard et al. (1951) and Elliott et al. (1973).

Figure 2-1 is a suggested flow diagram for analytical techniques that measure C_{11} plus hydrocarbons in waters, based on CCl_4 extraction. The latter section of Table 2-2 presents information about these methods.

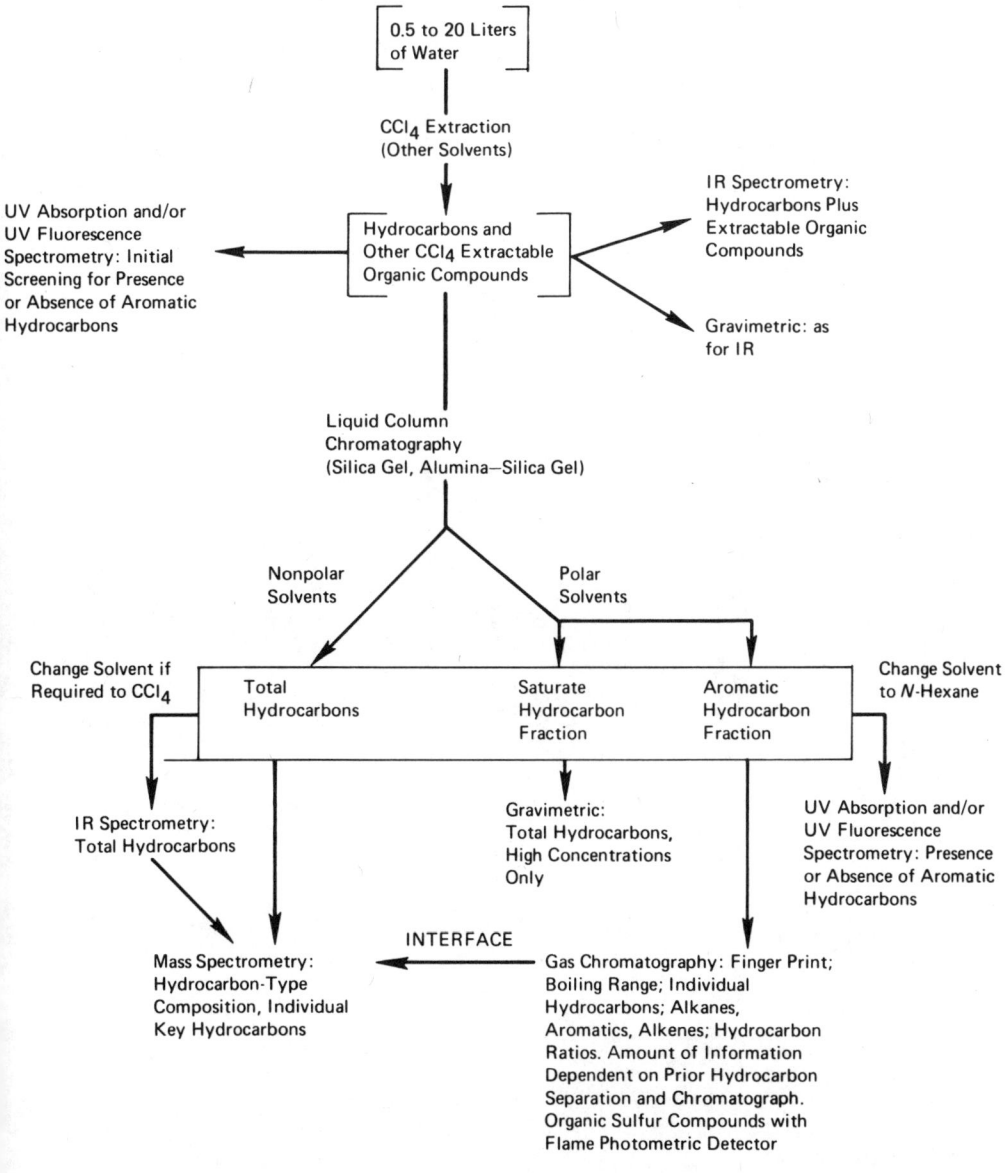

FIGURE 2-1 Flow diagram for analytical techniques to measure C_{11}^+ hydrocarbons in waters.

TABLE 2-2 Analytical Techniques for the Determination of Hydrocarbons in Water[a]

Technique	Component Determined	Sample Size (g)	Advantages	Disadvantages	Equipment (Approximate Cost in Dollars)	Analysis Time: Operator (Elapsed)	References
C_1–C_{10} *Hydrocarbons*							
Gas equilibration	Individual hydrocarbons and hydrocarbon type	50–250 ml	Parts per trillion sensitivity, separates hydrocarbons from nonhydrocarbons. No sample preparation	Analysis time relatively long	Gas chromatographs (15,000)	10–30 min (0.5–2 h)	McAuliffe, 1969, 1971
Gas stripping	Individual hydrocarbons and hydrocarbon type	1–2 liters	Measure background levels of C_1, C_2, C_2^-, C_3, C_3^-, i-, n-C_4 in open ocean waters	Nonhydrocarbons can interfere. Analysis time relatively long	Gas chromatograph (15,000)	10–30 min (0.5–1 h)	Swinnerton and Linnerbom, 1967
Vacuum degassing	Individual hydrocarbons and hydrocarbon type	4–20 liters	As above, can be used to continuously measure hydrocarbons in water	Normally used to measure C_1–C_4 in seawater. Equipment expensive	Complete system—pumps, gas chromatograph (300,000; 2,500 per day rental)	3–30 min (3–30 min)	Schink et al., 1971; Fort et al., 1973
C_{11} *plus Hydrocarbons*							
Gravimetric	Nonvolatile extractables	1–4 liters	Simple, minimum equipment	Nondiagnostic, conc. between 0.3–1,000 mg/liter	Glassware, balance (1,000)	20 min (40 min)	Environmental Protection Agency, 1971
UV absorption spectrometry	Conjugated polyalkenes, aromatics	1 liter	Useful for conc. > 10 μg/liter	Not very diagnostic, less sensitive than fluorescence spectrometry. No information on saturated HCs	UV absorption spectrometer (3,000–5,000)	20 min (20 min)	Levy, 1971
UV fluorescence spectrometry	Unsaturated compounds, aromatics	1 liter	Useful for conc. > 10 μg/liter; measure HCs in open ocean waters	Not very diagnostic. No information on saturated HCs. Fluorescence may be quenched	Fluorescence spectrometer (10,000)	20 min (20 min)	Levy, 1971; Thurston and Knight, 1971; Zitko and Carson, 1970
Infrared spectrometry	Methyl, methylene, carbonyl, aromatic. Total hydrocarbons	1–4 liters	Information on functional groups. Identify contaminants such as silicones, plasticizers	Concentrations >3 μg/liter, 0.1 mg. Not very diagnostic	Low or high resolution infrared spectrometers (3,500–35,000)	5–10 min (20–40 min after separation from water)	Brown et al., 1973; Kawahara, 1969; Simard et al., 1951
Gas chromatography (low resolution)	Hydrocarbon profiles and boiling range of sample, C_{11}–C_{50}	1–20 liters	Quick examination, reasonably diagnostic	Little information from highly weathered or biodegraded oils	Gas chromatography (10,000–15,000)	10 min (2 h)	Adlard et al., 1972; Brown et al., 1971; Duckworth, 1971; Ehrhardt and Blumer, 1972
Gas chromatography (high resolution), special detectors	More detailed hydrocarbon profiles. Sulfur profiles, individual hydrocarbon ratios, C_{11}–C_{40}	1–20 liters	Better diagnostic power. Sulfur compounds assist in identification	Little information from highly weathered or biodegraded oils	Gas chromatograph (15,000–20,000)	10 min (2 h)	Kreider, 1971; Miller, 1973; Ramsdale and Wilkinson, 1968; Zafiriou et al., 1973
Mass spectrometry	Hydrocarbon types	1–10 liters	Provides complete HC type information	Complex and expensive equipment. Requires computer interface	Low resolution, mass spectrometers (60,000). High resolution, (80,000–150,000)	10 min (2 h)	Aczel et al., 1970; Hastings et al., 1956; Hood and O'Neal, 1959; Robinson, 1971
Gas chromatograph, mass spectrometer	Specific hydrocarbon, C_4–C_{30}	1–10 liters	Identify and measure individual hydrocarbons	Very complex and expensive equipment	Add gas chromatograph cost to above	10 min (2 h) 2–4 h (2–4 h)	

[a] Hydrocarbons are extracted from water and then separated from nonhydrocarbons by column or thin layer chromatography.

Gravimetric Methods

The literature describes several standard gravimetric methods for quantitatively determining the relatively high concentrations of oils and greases in discharges from cities, ships, etc. The extractable material is separated with the solvent, which is then evaporated, and the residue is weighed (American Public Health Association, 1971; Environmental Protection Agency, 1971; American Society for Testing Materials, 1972). The methods measure extractable amounts from 5 to 1,000 mg/liter.

Separation of nonhydrocarbon compounds by column chromatography, using hexane as a solvent, was recently proposed as a standard method for determining petroleum in seawater (Soviet Submission to IMCO, 1973). The lower limit of detection of the method, however, is only 0.3–0.5 mg/liter—far in excess of that normally encountered in ocean waters.

UV Absorption Spectrometry

Carbon tetrachloride extracts of 1-liter samples of seawater are evaporated to dryness, and the residues are dissolved in normal hexane (Levy, 1970). The UV absorption spectra of these solutions are scanned from 350 to 210 nm, and their absorbances at 256 nm are compared with those of a series of standard solutions prepared from the source oil.

The method provides a quantitative measure of the amount of oil dissolved or finely dispersed in the water column. The method is suitable if concentrations exceed 10 μg/liter. This procedure has been used onboard ship and was used to study at least one pollution incident (Levy, 1970). Ultraviolet absorption spectrometry is relatively nonspecific; any substance that possesses the necessary solubility characteristics and absorbs light at 256 nm will be interpreted as being from a petroleum source.

UV Fluorescence Spectrometry

Fluorescence spectrometry has proved useful in several investigations of petroleum-derived hydrocarbons in the marine environment (Levy, 1971, 1972; Levy and Walton, 1973). The method is more specific than absorption spectrometry. The CCl$_4$ extract of the water sample is evaporated slowly to dryness and the residue dissolved quantitatively in normal hexane. Both the excitation and emission spectra for this extract are scanned with a fluorescence spectrometer. Fluorescing substances in crude and residual fuel oils are excited most strongly at 310 nm. The fluorescence intensity is compared with those of three standard solutions covering concentration ranges of a reference oil. The concentration of the unknown is obtained by interpolation, to within 1 μg/liter of petroleum ±10 percent precision.

The absolute accuracy of the method must await standardization of a reference oil; until then, results must be expressed in terms of some arbitrarily chosen standard. Because the petroleum-derived fluorescence results from many different oils, all of which may have been changed by weathering, the amount of oil present can only be estimated. Confirmation must be by other techniques, some of which follow.

Liquid Column Chromatography

Uncertainties introduced by nonhydrocarbons in the CCl$_4$ extract can be avoided by separating them from the hydrocarbons. The total hydrocarbons or fractions (types) can then be obtained by infrared, gas chromatography, mass spectrometry, or combinations of these techniques. Thin-layer chromatography will work in principle, but it is less precise and less convenient to use than others available. This discussion will be restricted, therefore, to column chromatography.

Activated silica gel has been used (Brown et al., 1973). When CCl$_4$ extract has been evaporated to 2 ml, 0.2 ml of n-pentane is added. The hydrocarbons are eluted with a CCl$_4$, n-pentane mixture to obtain a saturate fraction. Elution with CHCl$_3$ gives an aromatic fraction.

A combined alumina-silica gel column has also been used (Blumer et al., 1970; Ehrhardt, 1972). The column is deactivated with 5 percent water prior to packing in order to avoid hydrocarbon artifacts. The CCl$_4$ extract is changed to n-pentane. The 2-ml pentane extract is placed on the alumina-silica gel column and eluted with n-pentane. A total hydrocarbon fraction is obtained, without separation between saturate and aromatic hydrocarbons.

The hydrocarbon fractions can then be analyzed by the techniques shown in Figure 2-1.

Infrared Spectrometry

Infrared is commonly used for analyzing properties of hydrocarbons, such as determining the low levels of hydrocarbons present in open ocean waters (Brown et al., 1973). In Brown's analysis, carefully collected water samples were repeatedly extracted with high-purity CCl$_4$. The extract was concentrated by evaporation and examined by IR to measure total extractable organic compounds (hydrocarbons as well as organic compounds such as fatty alcohols and acids). Hydrocarbons were then isolated by activated silica gel chromatography. High sensitivity was obtained by paying careful attention to sources of contamination. A high-resolution IR spectrometer with scale expansion and a long path length cell can determine hydrocarbons to approximately 1 ppb in a 6-liter water sample.

Gas Chromatography

The high discrimination obtainable by gas chromatography is described in the work of Zafiriou et al. (1973). He used a support-coated open tubular (SCOT) column of 50 ft × 0.02 in., coated with OV-101 and rated at 25,000 effective plates. In an alternate approach Kreider (1971) used a 10 ft × 0.12 in. packed column with 10 percent OV-101 on 80/100-mesh chromosorb W. British Institute of Petroleum subpanel ST-G-GD recommends a two-level approach—a first look at the as-received sample with a low-resolution packed column and examination of less than 340 °C material with a 50 m × 0.5 mm stainless steel capillary column coated with OV-101 and programed from 100 to 200 °C at 5 °C/min. Other packings used include nonpolar phases such as Apiezon L, and SE-30, and polar phases such as FFAP.

The simplest form of gas chromatography, the chromatogram, can act as a fingerprint, providing information on boiling range. More detailed data, especially that from open tubular columns with high resolution, can provide substantial information on individual alkane hydrocarbons, which can indicate the origin of the materials. Ehrhardt and Blumer (1972) describe the detailed analysis of oils to distinguish commercial petroleum hydrocarbons from recent biogenic hydrocarbons.

Chromatograms of the hydrocarbons permit calculation of odd–even *n*-alkane ratios and *n*-alkane–isoprenoid ratios. Such analyses are best measured on a saturate fraction obtained by liquid chromatography.

Additional discriminating power can be obtained with selective detectors. Adlard et al. (1972) recommended the use of the flame photometric sulfur detector, and this has been reported by Miller (1973) and others.

If sufficient hydrocarbon is present in the solvent, only a small portion need be injected onto the chromatographic column. However, many samples of water, sediments, and tissue may have relatively low concentrations. To obtain adequate signal above the column bleed, it is necessary to inject a larger volume of the sample.

Excessive solvent produces a large interfering peak, and it is necessary to remove most of the solvent. Blumer et al. (1972) accomplished this by evaporating the sample onto the exterior of an etched glass tube that was then inserted into the injection port of the gas chromatograph. After a specified time, the tube may be removed to prevent the entrance of very high-molecular-weight hydrocarbons onto the chromatographic column. Ramsdale and Wilkinson (1968) evaporated the sample in the end of a small glass cup that encased an iron rod. The cup was inserted through a double-valve system into the gas chromatograph inlet, and after a predetermined period of time it was removed with a magnet. Perkin–Elmer's new sample injection system is excellent for introducing solvent-free samples.

Mass Spectrometry

Hydrocarbon composition can be obtained by mass spectrometry. For saturate fractions, a commonly used method is that of Hood and O'Neal (1959). Aromatic fractions can be most completely characterized by a low-voltage, high-resolution mass spectrometer (Aczel et al., 1970), but this requires approximately 5 mg of sample and is not practical for many environmental problems. Consequently, the more usual method is high-voltage operation at a resolution of 1 part in 500 (Hastings et al., 1956), which has been shown to be usable with as little as 6 μg (Brown et al., 1973). Smith (1972) has applied a method similar to Robinson's (1971) to a total hydrocarbon fraction collected from crude oil. This procedure determines alkanes, mono- through hexacycloalkanes, and eight aromatic types covering one to three aromatic rings per molecule. This method is of value for samples containing hydrocarbons at the microgram level, as it would eliminate the need to obtain separate saturate and aromatic fractions.

Gas Chromatography–Mass Spectrometry

Even high-resolution gas chromatographs (GC) do not completely resolve all hydrocarbon components. If a mass spectrometer is interfaced with the gas chromatograph, much greater diagnostic ability occurs. However, such a system is extremely expensive and is limited in the carbon-number range covered. It does have ability to obtain information on specific hydrocarbons.

Hydrocarbons in Biological Materials

Isolation of hydrocarbons from tissue samples differs from the procedures discussed above for waters, but once the hydrocarbons are separated, the analysis scheme is the same. Only techniques specific to tissue samples will be discussed.

Sample Collection and Preservation

Care must be taken to avoid contamination of the organism. Small samples can be wrapped in clean aluminum foil, placed in glass jars, and immediately frozen and kept frozen until analyzed. Large organisms such as fish can be wrapped in clean foil and frozen.

EXTRACTION AND SAPONIFICATION

Several methods have been used to isolate hydrocarbons from organisms. Extraction with organic solvents in a Soxhlet apparatus or using homogenization have been described (e.g., Farrington et al., 1972; Mackie et al., 1972). Connell (1971) extracted samples of fish with diethyl ether, followed by steam distillation of the ether extract to obtain a volatile fraction. Some analysts have advocated digestion of the sample in alcoholic KOH, followed by partitioning of the hydrocarbons and unsaponificable lipids into a nonpolar solvent such as hexane (Greffard and Meury, 1967; Blaylock et al., 1973). The advantages are said to be disruption of cells and more efficient extraction. A comparison is needed of the efficiencies and selectivity of the various extraction procedures.

Most extracts obtained by the above methods contain esters of fatty acids that often interfere with the isolation of alkene and aromatic hydrocarbons. Saponification will break the esters into fatty acid salts and alcohols, which are easily removed. However, one must avoid transesterification of the existing esters to methyl or ethyl esters. Complete saponification is confirmed by the absence of the carbonyl absorption band for esters in the infrared spectra of the material remaining after saponification.

SEPARATION OF HYDROCARBONS FROM LIPIDS

A variety of techniques have been used to separate hydrocarbons from coextracted lipids. Thin-layer and column chromatography have been used, singly and in combination (Ehrhardt, 1972). High-pressure liquid chromatography should gain wider acceptance after further research and testing. Karger et al. (1973) used 2,4,7-trinitrofluorenone-impregnated Corasil I high-pressure liquid chromatography columns to separate isomers of polynuclear aromatic hydrocarbons.

ANALYTICAL METHODS

The following techniques are reviewed to identify those that will best assist the analyst in detecting petroleum hydrocarbons in the presence of comparable or higher levels of biogenic hydrocarbons.

Infrared Spectrometry

The absorption frequencies of biogenic hydrocarbons overlap or coincide with those for petroleum hydrocarbons (with the possible exception of aromatic hydrocarbons in the long wavelength region). Thus, this technique shows little promise for detecting small quantities of petroleum hydrocarbons in the presence of biogenic hydrocarbons.

UV Absorption and UV Fluorescence Spectrometry

These analytical techniques detect aromatic hydrocarbons (Thurston and Knight, 1971; Zitko and Carson, 1970). They can be applied to lipid extracts and unsaponifiable lipid extracts without isolating the hydrocarbons. Appropriate attention must be directed toward the possible presence of biogenic aromatic hydrocarbons and the quenching of fluorescence emission. These techniques do not give an indication of the complexity or molecular weight range of the mixture.

There is possible overlap of the UV absorption spectra of highly conjugated biogenic alkenes and aromatic hydrocarbons. The methods do not detect alkanes or nonconjugated alkenes.

Gas Chromatography

Gas chromatography is commonly used and provides considerable information. Some investigators have estimated petroleum hydrocarbons from the unresolved compounds that reside beneath the resolved compounds in the gas chromatogram (Farrington et al., 1972; Burns and Teal, 1971; Farrington and Quinn, 1973). This is subject to error because the amounts of resolved or partially resolved peaks sometimes are not included in the estimate. Also the size of the unresolved complex mixture relative to the resolved peaks is a function of the resolution of the column. For example, a high-resolution capillary column produces many more resolved peaks and a much smaller unresolved envelope than do medium or low-resolution pack columns. Thus, estimates will vary from laboratory to laboratory.

Normal alkanes from biogenic sources usually have a ratio of 5:3, odd- vs. even-carbon-number preference C_{25}–C_{33} range, whereas those from many crude oils have little or no preference. Some crude oils do show a slight odd-carbon number preference, but seldom above 1.2. An odd-to-even n-alkane ratio of one in the C_{25}–C_{33} range may not always be crude oil if found, because a few bacteria have been reported to generate alkanes with this distribution (Davis, 1968), as well as from sponges and corals (Koons et al., 1965). Other carbon number ranges such as C_{15}–C_{19} may show differences between these organisms and crude oil.

The amounts of C_{16} through C_{20} isoprenoid isoalkanes in crude oils or refined products may be useful in distinguishing petroleum from biogenic hydrocarbons. Marine organisms have been shown to produce pristane

(C_{19}) in appreciable concentrations, but very little if any of the other isoprenoids in the C_{16}–C_{20} range. Many petroleums contain appreciable amounts of C_{16}, C_{18}, pristane (C_{19}) and phytane (C_{20}), but low amounts of the C_{17} homolog.

In many crude oils the ratio of pristane to phytane falls between 1.5 : 1 and 2.5 : 1. This and the pristane to C_{18} isoprenoid isoalkane may be diagnostic. Blumer and Sass (1972) have suggested this means for distinguishing petroleum and biogenic hydrocarbons in sediments. Organisms containing only biogenic hydrocarbons should have extremely high ratios (>100 : 1); petroleum hydrocarbons, about 2 : 1.

Clark and Finley (1973) described the use of the ratio of n-C_{16} to the sum of the normal alkanes C_{14}–C_{37}. This ratio tends to be smaller (<15 : 1) for petroleum samples than for biological materials.

Some marine organisms contain alkane hydrocarbons with one or two predominating, which is not true for either crude oils and refined products. However, this identification is not positive.

Gas chromatography methods that employ the quantification of one or a series of resolved petroleum hydrocarbon peaks as a means of estimating the concentration of petroleum in an organism are subject to error. These methods require the determination of the ratio between the resolved component(s) and unresolved components in the crude oil or fuel oil that was the source of the petroleum in the sample in order to convert from resolved component(s) concentration to total petroleum concentration in the sample. It is assumed that the ratio of the resolved components to total petroleum is not altered by weathering or by incorporation into the organisms, an assumption of questionable validity (Blumer et al., 1970).

Also, many of these methods are only useful in establishing whether petroleum is present in a sample and do not attempt to measure the amount of oil.

Brown et al. (1971) developed a GC/UV method to measure eighteen polynuclear aromatic compounds. A sample is tagged with ^{14}C—benz[a]anthracene and benzo[a]pyrene, treated with aqueous caustic, and chromatographed on partially deactivated alumina. A portion of the resulting polynuclear aromatics (PNA) concentrate is injected into a gas chromatograph; individual peaks trapped and then quantified by UV absorption spectrophotometry. The method has been used to measure these compounds in tars from automobile exhaust gas, but can be applied to crude oils.

Because marine tissues often contain no more than 100 ppm total hydrocarbons, impractically large amounts of tissue would be needed to supply the 1 μg of each polynuclear aromatic hydrocarbon required for measurement. The most pertinent nongas chromatographic methods for tissue are described by Howard et al. (1968) and Grimmer and Hildebrant (1972). Individual PNAs are measured down to 2 ppb in 200 g samples of dairy products, meat, fish, vegetables, and beverages.

Mass Spectrometry

Mass spectrometry and gas chromatography–mass spectrometry can indicate the amount of petroleum hydrocarbons in a sample. Hydrocarbons isolated from shellfish and sediments containing petroleum have shown fragments of the series of alkylated aromatic hydrocarbons of a distribution and complexity similar to that found in petroleum (Ehrhardt, 1972; Tissier and Oudin, 1973). Brown et al. (1973) used similar techniques in analyzing hydrocarbons in seawater.

We believe that careful selection and application of the techniques suggested above will detect the presence of petroleum hydrocarbons in marine organisms in most cases where the ratio of petroleum hydrocarbons to natural background of hydrocarbons is 1.0 or greater.

Hydrocarbon Analysis of Sediments

Sample Collection and Preservation

In water depths up to 100 feet, hand coring by scuba divers is recommended. Unlike gravity coring, the substrate maintains integrity, and there is little compaction of the sample. Alternatively, or in deeper waters, gravity or box cores are best, or various dredges such as the Peterson dredge; subsequently, these deeper samples can be "cored."

Analytical Methods

Low-Molecular-Weight Hydrocarbons (C_1–C_{10})

The frozen core sample (sealed) should be dipped in boiling water to thaw the tube and surface of the core. The material can then be extruded and a portion cut from the top. This is divided to analyze for low- and high-molecular-weight hydrocarbons.

The sample for low-molecular-weight analysis is transferred to a 125-ml, narrow-mouth glass bottle and covered with hydrocarbonfree water, leaving a 30-ml gas space. The air is displaced with nitrogen or helium in a dry box, and the bottle is capped with a Teflon-lined metal screw cap. The sample is analyzed as described for water.

Mommessin et al. (1968) measured C_4 through C_7 hydrocarbons by grinding the sample in a sealed brass cylinder at 120 °C. Helium flushed through the cylinder was passed through traps to remove water, the hydrocarbons trapped in a cold trap, and later released to a gas chromatograph.

Brown et al. (1973) heated the samples and vaporized the hydrocarbons directly into the inlet system of a mass spectrometer.

High-Molecular-Weight Hydrocarbons (C_{11}^+)

The C_{11}^+ hydrocarbon content of unpolluted sediments varies from approximately 1 to 100 ppm, although high values are found in oil seep areas such as Coal Oil Point, Santa Barbara Channel. The lower concentrations, in particular, suggest the need for careful evaluation of any proposed method, with regard to sample recovery and contamination from solvents or other sources.

Many workers have reported measurements of hydrocarbons in sediments and soils. Analysis generally consists of drying and extracting the sample, separating hydrocarbons from extracted nonhydrocarbons, and hydrocarbon measurement and characterization.

After drying, the sample is usually ground or powdered, to let the extracting solvent contact the sediment completely. Solvent extraction procedures vary widely (Brown et al., 1973). Soxhlet apparatus with hot solvent is often used, as well as repeated extraction with hexane or CCl_4 at room temperature imparting energy by shaking or with ultrasonics.

It is important to remove elemental sulfur from sediment extracts prior to hydrocarbon analysis, to avoid interference (Blumer, 1967; Farrington et al., 1972). The sediment extract can then be analyzed as suggested in the flow diagram of Figure 2-1.

The criteria for distinguishing between petroleum and biogenic hydrocarbons in sediments are the same as described for analysis of water and organisms.

SUMMARY AND RECOMMENDATIONS

A number of analytical techniques are available for measuring low- and high-molecular-weight and total hydrocarbons in samples of oil, water, sediment, and tissue. The methods vary widely in their detection limits and diagnostic abilities. They range from the measurement to a few parts per trillion (10^{12}) of individual low-molecular-weight hydrocarbons (C_1–C_{10}) in water to the nondiagnostic gravimetric determination of hydrocarbons in sediment and tissues with detection limits about one million times less (a few ppm).

As the complexity of petroleum hydrocarbons increases with molecular weight, the analysis for individual or even classes of hydrocarbons becomes more difficult. However, some techniques can partly characterize high-molecular-weight (C_{11}^+) hydrocarbons. These include high-resolution gas chromatography and combined gas chromatography–mass spectrometry.

Hydrocarbons of biogenic origin often differ from petroleums. Some of the analytical methods, such as gas chromatography following hydrocarbon-class-type separation by column chromatography, can often distinguish hydrocarbons from these two sources. This is more likely if petroleum hydrocarbons represent greater than 10–20 percent of the total hydrocarbons, and the total in the sample is not so low as to approach the limits of detection.

The analytical techniques reported here have been restricted to the determination of hydrocarbons. Only brief mention is made of carbon, sulfur, nitrogen, and metals in oils. For proper evaluation many biological studies also require accurate monitoring of such other variables as temperature, oxygen, salinity, and nutrients.

We have identified several problem areas that require further development of methods:

• *Sample collection.* Do samplers contribute or remove hydrocarbons from the sample? What are the best sampler materials and how does one obtain, for example, subsurface water samples without contamination from hydrocarbons on the water surface or near surface?

• *Preservation of samples.* How is a sample best sealed to prevent loss of the volatile hydrocarbons? How is bacterial deterioration of hydrocarbons prevented? Is it sufficient to add mercuric chloride to aqueous samples, or store sediment and tissue samples at low temperature?

• *Very low concentration analysis.* The hydrocarbon contents of the open oceans are low, a few parts per billion. Current techniques can detect approximately 1 ppb or less. More sensitive methods and more diagnostic methods are required, to accurately measure these hydrocarbons and to determine the portions from petroleum and biogenic sources.

• *Separation of petroleum from other hydrocarbons.* Methods for distinguishing petroleum hydrocarbons from biogenic hydrocarbons in sediments, organisms, and water are needed.

Biological

This section outlines criteria that should be used in selecting techniques for assessing the biological effects of oil spills or of continuous oil additions to the marine environment. Better methods are needed for determining the biodegradation of oil, the bio-uptake of oil, and influences various oils have on marine organisms in the intertidal zone, benthos, and water column. In this section, biological damage is defined as "significant" only if it impairs the survival of species essential to the ecosystems.

Most marine organisms show natural seasonal variations that are related to yearly cycles, as well as year to year variations. Natural calamities in the marine environment can be caused by changes in salinity, temperature, oxygen level, and the buildup of poisonous materials or gasses. Phytoplankton are subject to rapid and drastic changes within a season, with one species of diatom or dinoflagellate taking over the predominant position held by another species. These drastic changes may be caused in part by changes in temperature, light, or the availability of nutrients (Korringa, 1973). These natural occurrences, causing variations in species composition, make it difficult to detect in the field changes caused by petroleum additions. If multiple natural occurrences coincide with an oil spill [such as occurred at Santa Barbara (Straughan, 1971)], separation of the effects of petroleum becomes difficult.

Because of the large natural variations in marine life, all of the important variables that may influence biological response should be monitored following an oil spill. These include weather, hydrodynamic conditions, temperatures, salinities, suspended particulate matter, and the oil contents of waters and sediments. Similar information should be available for chronic oil discharges or experimental studies. The number of hydrocarbon-oxidizing bacteria should also be determined because they too may influence interpretation of results.

In controlled experiments, proper statistical approaches must be used to obtain a measure of the variance, and the environmental parameters must be varied in a manner that will ensure the proper identification of damage-causing agents. The number of variables influencing marine life is large, and experimentation can become time-consuming and expensive. Well-designed partial factorial experiments can often establish the relative importance of various parameters.

MICROBIAL BIODEGRADATION

POPULATION ENUMERATION

Heterotrophic-bacteria populations may increase as oil becomes a carbon and energy source. Because of size differences of bacteria, population is not always proportional to biomass. These values depend not only on the method used but also on the number of small predators present and on substrate abundance.

Counting techniques are based on dividing a sample into an appropriately small volume so that separate organisms can be isolated and grown to a detectable level. Usual procedures involve variations of either the most probable number technique (Gunkel, 1968; Robertson et al., 1973) or the direct plate count (Buck and Cleverdon, 1960). The most probable number technique involves statistically determining the minimum raw seawater volume that can generate a response when inoculated into some receptive medium. Pertinent responses would include the ability to produce recognizable and measurable products from the added oil, such as cell tissue, carbon dioxide, and adenosine triphosphate (ATP).

In the course of a few months, in laboratory experiments, disruption of an oil slick is visible evidence that active oil-metabolizing organisms were present. Population numbers obtained depend on the procedure used. For instance, the populations of some species are measured by adding an emulsified solution to break up clumps of organisms. Although this method increases individual counts, it possibly kills some organisms. The results of selective population determinations (such as hydrocarbon oxidizers) by plate count at various dilutions are also difficult to interpret because some organisms do not thrive on solid media, while others use low-level contaminating organic substrates rather than that intended and supplied.

The submerged culture generation method is probably preferable because of its sensitivity, particularly when resulting populations are determined by their ATP content. However, present methods can only yield gross relative comparisons. In any event, because bacteria seem to occur throughout the ocean (Gunkel, 1968; Button et al., 1973), further refinement of these methods would seem unnecessary. However, these techniques may have importance in monitoring hydrocarbon-oxidizing bacteria in waters and sediments following oil spills or during laboratory experiments.

METABOLIC ACTIVITY

The general activity level of hydrocarbon oxidizers can be assessed by adding low concentrations of high-specific-activity ^{14}C-labeled hydrocarbons to raw seawater. After suitable incubation (ranging from hours to months) at ambient seawater temperatures and pressures, the amount of radioactive CO_2 produced from the radioactive hydrocarbon provides an index of the rate of oxidation of the labeled compound. Typically, microgram quantities of some hydrocarbons oxidize in days or weeks (Button, 1972). Such experiments give information only about individual petroleum compounds, and it must be remembered that petroleum composition is exceedingly complex and variable.

HYDROCARBON BIO-UPTAKE

Studies of the uptake, metabolism, storage, and discharge of hydrocarbons by various members of the marine food chain are important. Labeling hydrocarbons as radioactive or stable is a straightforward diagnostic technique for measuring the uptake, metabolism, and discharge of hydrocarbons from selected marine organisms. The validity of this technique has been amply demonstrated by the work of Lee et al. (1972a, 1973). Incorporation of various labeled hydrocarbons added in solution to the controlled habitat of various species of phytoplankton, zooplankton, fish, and benthic invertebrates has been studied. Labeling allows an accurate measurement of hydrocarbon uptake by an organism. This method can also establish the distribution of labeling retained in various cellular entities or excreted in the form of metabolized or unaltered hydrocarbons.

ALGAL RESPONSES

PHYTOPLANKTON

Phytoplankton is responsible for the fixation of the energy utilized by marine ecosystems. How oil in the water column affects their growth rates must be determined for both spills and continuous additions in estuarine, coastal, and oceanic environments.

Field Studies

The first step in evaluating the effects of oil on phytoplankton should involve field studies. One of these should determine the concentrations of oil in the water of the region of interest. Another should determine if the observed concentrations have a measurable effect on the growth of natural phytoplankton communities. Two techniques appear useful.

(1) Radiocarbon bicarbonate uptake has been employed by Dickman (1971), Strand et al. (1971), and Gordon and Prouse (in press). It is relatively quick (individual experiments take no longer than one day) and can be conveniently used in estuarine or open-sea environments. It is based on the standard radiocarbon method for measuring planktonic primary production (Strickland and Parsons, 1965). A known amount of oil in seawater is added to raw seawater samples (previously passed through a 160-μm screen to remove large zooplankton). Samples and controls are inoculated with ^{14}C-bicarbonate and incubated at ambient light intensity and in situ temperature for about 6 h. The radiocarbon uptake of samples containing oil is compared with that of controls. After at least 12 experiments over the entire range of expected concentrations, the relative radiocarbon uptake can be plotted against concentration to determine effects.

Several steps in this technique require particular attention (Gordon and Prouse, 1973). The forms and concentrations of oil added to seawater samples must be as similar as possible to those encountered by phytoplankton in the environment. These authors describe a suitable standardized procedure. This procedure should be standard for all experiments.

The oil concentrations must be measured with the best available chemical methods [for example, fluorescence spectroscopy (Keizer and Gordon, 1973) or gas chromatography] before and after incubation, so that the concentration of oil available to the phytoplankton during the experiment is accurately measured. An 18-h conditioning period in subdued light is recommended, after the additions of seawater containing hydrocarbons, but before radiocarbon inoculation.

(2) In situ cylinder or field fencing—a method for isolating a vertical cylinder of water (Kauss et al., 1973; Kinsey, 1973)—is designed to investigate the possible effects of an oil spill. This method is of no apparent use in studying the effects of chronic additions. Observation

periods of a few hours to several months restrict its use to easily accessible and sheltered environments such as lakes, estuaries, coastal inlets, and back reefs.

Portions of the water column are isolated in open-ended plastic cylinders that extend from the surface to the bottom sediments. Oil is added to some; others serve as controls. Water is collected periodically by siphon from beneath the contained slicks, and phytoplankton concentrations are determined by microscopic examination, care being taken to collect representative samples. The water should also be analyzed for oil composition and concentration. The effects of the oil can be evaluated by comparing changes in the numbers of the individual phytoplankton species in the oil-contaminated and oil-free waters. Thus, the effects on both total population and individual species can be determined simultaneously.

This field-type study is an attempt to simulate conditions that are likely to occur during an oil spill. Because the oil is confined and not allowed to disperse freely, however, the simulation is not perfect. Further caution should be exercised to ensure that the amount of oil used is comparable with an actual spill.

Effects of oil on gas (oxygen and carbon dioxide) exchange between water and atmosphere and the effect of oil on other flora and fauna has been studied with this technique (Kinsey, 1973).

Laboratory Studies

Once the expected range of oil concentrations in the area under study has been shown to have measurable effects on the growth of natural phytoplankton, laboratory experiments should be conducted with cultures of selected phytoplankton species. Procedures for maintaining cultures and conducting growth experiments are well documented (Guillard, 1962). The American Society for Testing Materials (ASTM) method for diatom/pollutant bioassay should be consulted.

Single species experiments can show how different phytoplankton species respond to various concentrations of the same oil (Mironov and Lanskaya, 1969; Strand *et al.*, 1971; Nuzzi, 1973; Kauss *et al.*, 1973). They can also show how the species responds to individual components of whole oils. These experiments will therefore determine the most sensitive species and the most toxic oils.

Mixed species experiments, on the other hand, simulate changes in species composition of natural phytoplankton populations exposed to various oil concentrations and compositions.

In laboratory studies, the most accurate technique for measuring cell growth is direct microscopic counts. In some instances, this should be supplemented with radiocarbon uptake and measurements of particle-size distribution and concentration by electronic counters. As in field studies, standardized preparations of oil-contaminated seawater must accompany adequate monitoring of oil concentrations and distributions. The concentrations employed must not exceed those that are measured in actual spills.

ATTACHED MACROSCOPIC ALGAE

Attached macroscopic algae can account for half of the total primary production in some coastal regions. They are particularly vulnerable to oil spills. Field studies seem to have been restricted to long-term observations in areas affected by spills as in Chedabucto Bay (Thomas, 1973) and the *Torrey Canyon* spill. Laboratory studies using radiocarbon bicarbonate uptake (Gordon and Prouse, in press) could provide valuable information in a relatively short time.

BENTHIC MICROFLORA

Because oil from spills can persist longest in sediments (Blumer and Sass, 1972), the effects of oil in benthic microflora should be examined. A technique that can determine probable shifts in diatom community structure (Patrick, 1949) can be used.

SEDENTARY INVERTEBRATE POPULATIONS

Studies should include sensitive species of the ecosystem that are available for systematic quantitative sampling. Emphasis should be placed on sedentary organisms, particularly subtidal benthic infauna, because these species escape the spill and because they show less natural variation in numbers than intertidal species (Day *et al.*, 1970; Lie, 1968; Sanders, 1960).

SAMPLING BENTHOS

Following a spill, estimates of the distribution of oil on the surface and in the water column should be ascertained by samplings of both affected and unaffected areas. Weather and oceanographic conditions should be recorded, as should water temperatures and salinities.

Samples of oil should be taken on the surface as near the source as possible, as well as from the surface, water column, and sediments with increasing distance from the source. Chemical analyses of the oil and field samples should be obtained.

Previous baseline surveys should not be used as a reason to adopt obsolete methods or devote efforts to areas that are different from the spill area (Connell,

1971). To be useful, baseline surveys must be kept current with the best methods. Because such studies are not likely to have been made in most spill areas, efforts must usually be concentrated on doing a thorough postspill study of both affected and adjacent unaffected areas, including the progressive changes of the affected area.

If possible, sampling should start immediately after the spill, by counting living and dead individuals of each species (Sanders *et al.*, 1972). Because large numbers of organisms are needed, trawls or dredges should be used to collect samples for the first evaluation of possible effects. Either cores or grabs should be used to obtain quantitative samples of known bottom-surface area (normally 0.02–0.1 m^2) and volume (Nichols, 1973). Replicated samples of sediment for chemical analyses should be collected. In general, two smaller samples provide more information than one large one, but small samples may be less repeatable, owing to patchiness of the benthic organisms. All individuals of the taxa considered should be counted.

Screen sizes of 1 mm or larger show considerable variation between samples, in part, because the effective mesh size of the larger screens varies with amount of debris, tubes, etc., in the samples. Reish (1959) has compared the fauna retained on a graded series of sieves and found that a 1.4-mm mesh opening misses 25 percent of the species and 90 percent of the individuals. However, larger sample sizes, such as three combined 0.1 m^2 samples screened through a 1.2-mm mesh, are more representative of the macrobenthic organisms because "patchiness" introduces less error. Microfaunal benthos are best sampled by 0.3- or 0.25-mm screens. If bottom sediments contain a large proportion of mineral particles larger than 0.25 mm, separation of the benthic organisms using a 0.25-mm screen becomes difficult.

Enough stations are needed to provide a statistically significant representation of the major substratum types and to clearly indicate that the samples are representative.

Just as the composition of oil cannot be described from a few components, the complexity of an ecosystem cannot be described if only a few species populations are sampled. Rather than sacrifice accuracy to obtain larger coverage, each sample should be as complete as possible, for several reasons:

1. The samples are more repeatable through time if all size classes of the taxa considered are counted;
2. Life histories of a number of species can be described;
3. Diversity assessments are more meaningful if whole taxa are considered (Hurlbert, 1971).

BENTHIC ANALYSIS

Samples should be stained with rose bengal and carefully sorted under a microscope (Sanders *et al.*, 1972). In all instances, species identification is essential for comparisons between areas. Cursory identification of species to the major groups should be made rapidly and eventually sorted to species by experts.

Samples may be grouped by using a number of indices of faunal similarity (Greig-Smith, 1964; Southwood, 1966). The stations may be ordered either in a trellis diagram of stations or clustered (Sokal and Sneath, 1963; Stephenson *et al.*, 1971; Williams, 1971). The resulting classification of samples by faunal similarity may then be compared with spatial and temporal sequences.

Biological effects are best described by following changes in the abundance and distribution of each species. Because such data are difficult to summarize, a variety of community indices are used. The simplest are number of species and number of individuals. Measures of species diversity are less sensitive to the absolute size of collections and have been shown to be particularly useful in pollution studies (Patrick, 1949; Wihlm and Dorris, 1966). A number of diversity indices have been proposed. The Shannon-Wiener index (1963) has been used perhaps more than others. It, however, is subject to "false low values," even with samples containing a large number of species and individuals if one species has a large number of individuals (greater than 30 percent of the total number of individuals).

INTERTIDAL STUDIES

Benthic intertidal organisms can be sampled and examined in the same manner as subtidal benthic species. Other intertidal flora and fauna can be qualitatively examined during low tides. A transect can be established from the splash zone to the water's edge. Square-meter areas selected along the transect should provide species identification of plants and animals and estimates of the numbers of individuals. Careful color photography is useful for documenting selected square meters, as well as overall views of the intertidal zone along a marked transect.

Examination of these selected transects four times a year, for several years, establishes seasonal and annual variations. An estimate of possible damage from an oil spill may be obtained by comparing similar oil-contaminated and uncontaminated areas. Such areas may be empirically similar to comparable geographical sites, though most likely not identical. Therefore, the intertidal zone should be studied sufficiently long to establish patterns of change following a spill.

The effect of oil on saltwater marsh vegetation can be studied by observing oil-treated versus nontreated areas in a uniform salt marsh (Baker, 1973). Salt or inland marshes and bays where oil is continuously discharged can be studied to establish the survival of vegetation; such as in some areas of the Gulf of Mexico coast.

Scuba diving can be used to qualitatively evaluate the macroflora and -fauna of the tidal zone to 30 m. If the water is sufficiently clear, visual observations can be documented by photography of marked areas.

EXPERIMENTAL TOXICITY ON INVERTEBRATES AND FISH

Assessment of damage to larger marine organisms requires comparing analyses made within several disciplines (La Roche, 1973; La Roche et al., 1973). All exposed and control organisms, regardless of exposure time, should be compared through definitive analytical procedures. These range from relatively unsophisticated static bioassays to flow-through bioassay systems to behavioral studies and, finally, to complicated histological or enzymatic studies.

Oil and oxygen concentrations, pH, number of hydrocarbon-oxidizing bacteria, and other variables that can affect the test organisms should be monitored throughout the test period for both static and flow-through toxicity bioassays.

ACUTE BIOASSAYS

Acute bioassays are normally conducted from 12 to 96 h, depending on the organism under study. Key species of a desirable ecosystem are selected, from which Tl_{50} values can be derived. The investigator should look for early signs of behavioral or functional anomalies that might prove valuable in interpreting the biological mechanisms affected.

Additional information can also be obtained by keeping surviving specimens of the Tl_{50} experiments with the control organisms in uncontaminated water. Postexposure changes and death rates can lead to further characterization of impaired vital functions.

Static Acute Bioassays

Procedures for static acute bioassays with oils have been standardized (LaRoche et al., 1970) and should be performed as described in *Standard Methods for the Examination of Water and Wastewater* (American Public Health Association, 1971). In addition, the investigation should use a standard seawater media, reproducible shaking or mixing for oil-seawater preparations, and a reference toxicant.

Chronic Bioassays

Chronic bioassays refer primarily to exposures greater than 96 h or to short repeated exposures. Both continued and pulsed exposures permit the survival of the experimental species. To identify the indices of significant impairment, systematic analysis of biological functions and responses should be performed.

Chronic bioassays are of two general types, depending on whether the water is *recirculated* or allowed to *flow through* only once. In both cases, particular attention should be paid to the quality of the water to which oil is added in measured amounts. Also, experimental designs must include an adequate number of control organisms to permit essential comparisons (La Roche, 1973).

The recirculating water system, although comparatively simple, requires a very large water supply per unit weight organism to maintain water quality. Also, both water and pollutant must be metered when mixed and monitored to ensure that exposures are at the desired chemical concentration. The method can sometimes be used for simple chemical pollutants. However, it may alter many chemicals, with time, to more or less toxic by-products (Glass, 1973). Neither the production of more toxic by-products from the parent pollutant nor the constant neutralization of toxicants are desirable alternatives.

Another difficulty with recirculating systems is the accumulation of toxic excretory products or food residues. Filtration and "foaming out" of residues have had only limited success because they only prolong the time between inevitable changes of water. Differential removal of excretion products and the pollutant may cause a complexity of interpretation of the results. For these reasons, recirculation systems should be used with caution. The preferred system—*flow-through water systems*—is plagued with supplying suitable water (i.e., free from pollutants). Organisms must be selected with care to ensure that they can survive the austerity of this artificial environment in the absence of pollutant. Also, it is extremely important to monitor the inflowing natural or artificial water for possible changes in composition.

In short, no system is entirely free of side effects and without limitations. However, the flow-through system is, at present, the best available.

Field Experiments

Field experiments involving the transfer of organisms between oil and oil-free areas can be performed. However, inherent difficulties are always associated with the transfer of organisms from a viable environment.

Repopulation of benthic areas can be attempted in oil-contaminated and oil-free environments to assess the degree of recovery.

Organisms from either the acute or chronic bioassays can be observed for behavioral effects or examined for other biological changes.

BEHAVIORAL STUDIES

Behavioral studies might comprise the following:

- Gross behavioral changes following exposure (e.g., loss of the ability to school for the silverside);
- Changes in discriminating abilities (e.g., inability to choose a normally desired environmental temperature, or unusual slowness in obtaining offered food);
- Behavioral responses to positive stimulation (e.g., loss of ability to perform a simple task for food);
- Other chemotactic phenomena (e.g., mating responses).

OTHER BIOLOGICAL CHANGES

More detailed studies to determine possible effects of oil on organisms might include:

- Gross anatomical changes (e.g., pigmentation);
- Increased or decreased oxygen consumption;
- Swimming endurance under various regimens of dissolved oxygen (Voyer and Morrison, 1971);
- Growth rate based on timed ^{14}C-leucine incorporation into tissue protein (Jackim and La Roche, 1973);
- Histological effects that would include (1) classical histopathological examinations (Gardner and La Roche, 1973), (2) histochemical evaluations, (3) electron microscope studies, and (4) such hematological observations as differential blood counts (Gardner and Yevich, 1970);
- Resilience of blood figured elements in responding to biophysical treatments (Curby, 1973; Palek et al., 1972);
- Enzymic studies, including the comparative evaluation of various systems (Jackim et al., 1970) from organisms exposed to specific pollutants (often referred to as in vivo studies) and from tissues exposed to pollutants in vitro.

In all instances, the objectives of these approaches are understood to establish that the detectable differences between control and exposed organisms are significant and impair or block survival under these conditions.

Experimental work on the toxicity of spilled oils must always be related to natural conditions. Accidental oil spills should thus be studied and monitored in situ, to determine conditions during and after the incident. Controlled experimental work can then establish the relevance and contributions of the observed variables by evaluating the associated fundamental biological changes. Determining toxicity criteria, for instance, requires controlled experiments. Through experimental work, variables can be identified and evaluated one at a time; only then can field studies define the problem associated with actual spills.

Field studies at the time of oil spills can be made where other pollutants might be present, providing the effects of these other pollutants are properly identified and evaluated. Additional field information on species distributions, densities, and diversities can be obtained where oil is continuously discharged, such as in oil-producing areas or active oil-seep areas, such as offshore Coal Oil Point in the Santa Barbara Channel. Nonspill field studies should not be complicated by making them in estuaries or coastal waters heavily contaminated with industrial or municipal pollutants.

Field observations must always be complemented by controlled experimental work to ensure validity of interpretation.

SUMMARY AND RECOMMENDATIONS

Measuring the effects of oil on marine life is difficult. Each experimental study must include an adequate number of controls whereby single variables are evaluated through interdisciplinary approaches so that the effects of different biological parameters can be resolved. Many earlier studies of the effects of oil cannot be adequately evaluated because (1) the experimental work was not properly designed (for example, lack of adequate replication) and (2) the oil concentrations and other variables that affect marine life were inadequately monitored.

Well-designed laboratory studies are essential in determining the precise effects of oils on selected marine organisms. Studies are needed to measure the range of oil concentration in chronic and spill-type situations. Laboratory studies are also needed to compare organisms. However, evaluations of effects must be made under adequately monitored field conditions.

Scientists working in this area should become acquainted with the complexities of biological studies and make every effort to design and conduct truly diagnostic experiments. Such studies should be interdisciplinary, including biologists, chemists, statisticians, physical oceanographers, meteorologists, and geologists. Such studies are expensive but should obtain useful results at less cost than many isolated investigations.

References

Aczel, T., D. E. Allen, J. H. Harding, and E. A. Knipp. 1970. Computer techniques for quantitative high resolution: mass spectral analyses of complex hydrocarbon mixtures. Anal. Chem. 42:341–347.

Adlard, E. R., L. F. Creaser, and P. H. D. Matthews. 1972. Identification of hydrocarbon pollutants on seas and beaches by gas chromatography. Anal. Chem. 44:64.

Albaugh, E. W., and P. C. Talarico. 1972. Identification and characterization of petroleum and petroleum products by gel permeation chromatography with multiple detectors. J. Chromatogr. 74:233–253.

American Public Health Association. 1971. Standard methods for the examination of water and wastewater. 13th ed. pp. 254–256. American Public Health Association, Washington, D.C.

American Society for Testing and Materials. 1972. Water: Atmospheric analysis. Part 23. Method D-2778, 769. ASTM, Philadelphia.

Andelman, J. B., and M. J. Suess. 1970. Polynuclear aromatic hydrocarbons in the water environment. Bull. W.H.O. 43:479–508.

Baker, J. M. 1973. Biological effects of refinery effluents, pp. 715–724. In Proceedings, Joint Conference on Prevention and Control of Oil Spills. American Petroleum Institute, Washington, D.C.

Blaylock, J. W., P. W. O'Keefe, J. N. Roehan, and R. E. Wilding. 1973. Determination of n-alkane and methylnaptha-lene compounds in shellfish, pp. 173–177. In Proceedings, Joint Conference on Prevention and Control of Oil Spills. American Petroleum Institute, Washington, D.C.

Blumer, M. 1967. Hydrocarbons in the digestive tract and liver of a basking shark. Science 156:390.

Blumer, M., and J. Sass. 1972. The West Falmouth oil spill. II. Chemistry. Woods Hole Oceanographic Institution. Tech. Rep. 72-19. Unpublished manuscript.

Blumer, M., G. Souza, and J. Sass. 1970. Hydrocarbon pollution of edible shellfish by an oil spill. Mar. Biol. 5:195–202.

Blumer, M., R. R. L. Guillard, and T. Chase. 1971. Hydrocarbons of marine phytoplankton. Mar. Biol. 8(3):183–189.

Blumer, M., P. C. Blokker, E. G. Cowell, and D. G. Duckworth. 1972. Petroleum. In E. D. Goldberg, ed. A Guide to Marine Pollution. Gordon and Breach Science Publishers, New York.

Boylan, D. E., and B. W. Tripp. 1971. Determination of hydrocarbons in seawater extracts of crude oil fractions. Nature 230:44–47.

Brown, R. A., T. D. Searl, W. H. King, W. A. Dietz, and J. M. Kelliher. 1971. Rapid methods of analysis for trace quantities of polynuclear aromatic hydrocarbons in automobile exhaust gasoline and crankcase oil. CRC-APRAC Project Cape-12-68. Final report, December. U.S. Document #PB 219-025.

Brown, R. A., J. J. Elliott, and T. D. Searl. 1973. Measurement and characterization of nonvolatile hydrocarbons in ocean water. Unpublished manuscript.

Bryan, D. E., V. P. Guinn, R. P. Hackleman, and H. R. Lukens. 1970. Development of nuclear analytical techniques for oil slick identification. Phase I. Prepared for U.S. Atomic Energy Commission, January. Washington, D.C.

Buck, J. D., and R. C. Cleverdon. 1960. The spread plate as a method for the enumeration of marine bacteria. Limnol. Oceanogr. 5:78–80.

Burns, K., and J. Teal. 1971. Hydrocarbon incorporation into the salt marsh ecosystem from the West Falmouth oil spill. Woods Hole Oceanographic Institution Tech. Rep. 71-69. Unpublished manuscript.

Button, D. K. 1972. Hydrocarbon biodegradation kinetics, pp. 307–322. Background papers, workshop on inputs, fates, and effects of petroleum in the marine environment. National Academy of Sciences, Washington, D.C.

Button, D. K., S. S. Dunker, and M. L. Morse. 1973. J. Bacteriol. 113:599–611.

Clark, R. C., Jr. 1966. Saturated hydrocarbons in marine plants and sediments. M.S. thesis, Department of Geology and Geophysics, MIT, Cambridge, Mass.

Clark, R. C., Jr. and J. S. Finley. 1973. Techniques for analysis of data to assess oil spill effects in aquatic organisms, pp. 161–172. In Proceedings, Joint Conference on Prevention and Control of Oil Spills. American Petroleum Institute, Washington, D.C.

Clutter, D. R., L. Petrakis, R. L. Stenger, Jr., and R. K. Jensen. 1972. Nuclear magnetic resonance spectrometry of petroleum fractions. Anal. Chem. 44:1395–1405.

Connell, J. H. 1971. A guide to the proper method of investigation of the effects of oil pollution on marine organisms. Submission to the Royal Commission on Oil Exploitation on the Great Barrier Reef, Part III.

Curby, W. A. 1973. Relationships of electronic assays to erythrocyte morphology in *Fundulus heteroclitus*. In preparation.

Davis, J. B. 1968. Paraffin hydrocarbons in the sulfate-reducing bacterium *Desulfovibrio desulfuricans*. Chem. Geol. 3:155–160.

Day, J. H., J. G. Field, and M. P. Montgomery. 1970. The use of numerical methods to determine the distribution of the benthic fauna across the continental shelf of North Carolina. J. Anim. Ecol. 40(1):93–125.

Done, J. N., and W. K. Reid. 1970. A rapid method of identification and assessment of total crude oils and crude oil fractions by gel permeation chromatography. Sep. Sci. 5:825.

Duckworth, D. F. 1971. Aspects of petroleum pollutant analysis, pp. 165–185. In P. Hepple, ed. Water Pollution by Oil. Institute of Petroleum, London.

Dickman, M. 1971. Preliminary notes in changes in algal primary productivity following exposure to crude oil in the Canadian Arctic. Can. Field-Nat. 85:249–251.

Ehrhardt, M. 1972. Petroleum hydrocarbons in oysters from Galveston Bay. Environ. Pollut. 3:257–271.

Ehrhardt, M., and M. Blumer. 1972. The source identification of marine hydrocarbons by gas chromatography. Environ. Pollut. 3:179–194.

Eisler, R., G. R. Gardner, R. J. Henekey, G. LaRoche, E. F. Walsh, and P. P. Yevich. 1972. Acute toxicology of sodium nitrilotriacetic acid (NTA) and NTA-containing detergents to marine organisms. Water Res. 6:1009–1027.

Elliott, J. J., R. A. Brown, and T. D. Searl. 1973. An infrared spectrophotometric method for measurement of hydrocarbons in ocean water. Anal. Chem. Unpublished manuscript.

Environmental Protection Agency, Water Quality Office. 1971. Methods for chemical analysis of water and wastes. No. 217. Analytical Quality Control Laboratory, Cincinnati, Ohio.

Farrington, J. W., and J. G. Quinn. 1973. Petroleum hydrocarbons in Narragansett Bay. I. Survey of hydrocarbons in sediments and clams, *Mercenaria mercenaria*. Estuarine Coastal Mar. Sci. 1:71–79.

Farrington, J. W., C. S. Giam, G. R. Harvey, P. L. Parker, and J. Teal. 1972. Analytical techniques for selected organic compounds. *In* E. D. Goldberg, ed. Marine Pollution Monitoring: Strategies for a National Program. U.S. Department of Commerce, Washington, D.C.

Fort, E. R., B. O. Prescott, and A. Walters. 1973. Mapping hydrocarbon seepages in water-covered regions. U.S. Patent 3,747,405.

Gardner, G. R., and G. LaRoche. 1973. Copper induced lesions in estuarine teleosts. J. Fish. Res. Board Can. 30:363–368.

Gardner, G. R., and P. P. Yevich. 1970. Histological and hematological responses of an estuarine teleost to cadmium. J. Fish. Res. Board Can. 27:2185–2196.

Gerarde, H. W., and D. G. Gerarde. 1962. The ubiquitous hydrocarbons. J. Assoc. Food Drug Off. U.S. 25:26.

Glass, G. E., ed. 1973. Bioassay Techniques and Environmental Chemistry. Ann Arbor Science Publishers, Ann Arbor, Mich.

Gordon, D. C. W., and N. J. Prouse. The effects of three different oils on marine phytoplankton photosynthesis. Mar. Biol. In press.

Greffard, J., and J. Meury. 1967. Note sur la pollution en rode de toulou par les hydrocarbures canicerigenes. Cah. Oceanogr. 19(6):457–468.

Greig-Smith, P. 1964. Quantitative Plant Ecology. Butterworths and Co., London.

Grimmer, G., and A. Hildebrandt. 1972. Concentration and estimation of 14 polycyclic aromatic hydrocarbons at low levels in high protein foods, oils and fats. J.A.O.A.C. 55:631–635.

Guillard, R. R. L. 1962. Salt and osmotic balance. *In* Physiology and Biochemistry of Algae. Academic Press, New York.

Gulf Universities Research Consortium. 1973. A handbook on procedures and methods employed in the offshore ecology investigation (OEI). Gulf University Research Consortium, Galveston, Tex.

Gunkel, W. 1968. Bacteriological investigations of oil-polluted sediments from the Cornish coast following the *Torrey Canyon* disaster, pp. 151–158. *In* E. B. Cowell, ed. Biological Effects of Oil Pollution on Littoral Communities (Supplement to Field Studies 2).

Han, J., and M. Calvin. 1969. Hydrocarbon distribution of algae and bacteria, and microbiological activity of sediments. Proc. Natl. Acad. Sci. 64:436–455.

Hastings, S. H., B. H. Johnson, and H. E. Lumpkin. 1956. Analysis of the aromatic fraction of virgin gas oils by mass spectrometer. Anal. Chem. 28:1243.

Hites, R. A., and K. Biemann. 1972. Water pollution: Organic compounds in the Charles River, Boston. Science 178:158–160.

Hoefs, J. 1973. Stable Isotope Geochemistry. Springer-Verlag, New York.

Hood, A., and M. J. O'Neal. 1959. Status of application of mass spectrometry to heavy oil analysis. *In* J. D. Waldron, ed. Advances in Mass Spectroscopy. Pergamon Press, New York.

Howard, J. W., T. Fazio, R. H. White, and B. A. Klimeck. 1968. Extraction and estimation of polycyclic aromatic hydrocarbons in total diet composites. J.A.O.A.C. 51:122.

Hurlbert, S. H. 1971. The nonconcept of species diversity: A critique and alternative parameters. Ecology 52:577–586.

Institute of Petroleum Standardization Committee. 1973. Characterization of Oil Pollutants: Sampling, Analysis and Interpretation. Applied Science Publishers, New York.

Jackim, E. H., and G. LaRoche. 1973. Protein synthesis in *Fundulus heteroclitus*. J. Comp. Biochem. Physiol. 44A:851–866.

Jackim, E. H., J. M. Hamlin, and S. Sonis. 1970. Effects of metal poisoning on five liver enzymes in the killifish (*Fundulus heteroclitus*). J. Fish. Res. Board Can. 27:383–390.

Karger, B. L., M. Martin, J. Loheac, and G. Guiochon. 1973. Separation of polyaromatic hydrocarbons by liquid solid chromatography using 2,4,7-trinitrofluorescence impregnated Corasil I columns. Anal. Chem. 45:496–500.

Kauss, P., T. C. Hutchinson, C. Soto, J. Hellebust, and M. Griffiths. 1973. The toxicity of crude oil and its components to freshwater algae, pp. 703–714. *In* Proceedings, Conference on Prevention and Control of Oil Spills. American Petroleum Institute, Washington, D.C.

Kawahara, F. K. 1969. Laboratory guide for the identification of petroleum products. U.S. Environmental Protection Agency, Office of Research and Monitoring. National Environmental Research Center, Cincinnati, Ohio.

Keizer, P. D., and D. C. W. Gordon. 1973. Detection of trace amounts of oil in seawater by fluorescence spectroscopy. J. Fish. Res. Board Can. 30:1039–1046.

Kinsey, D. W. 1973. Small-scale experiments to determine the effects of crude oil films on gas exchange over the coral back-reef at Heron Island. Environ. Pollut. 4:167–182.

Koons, C. B., G. W. Jamieson, and L. S. Cierozsko. 1965. Normal alkane distributions in marine organisms. Possible significance to petroleum origin. Bull. Am. Assoc. Pet. Geol. 49:301–304.

Koons, C. B., P. H. Monaghan, and G. C. Bayliss. 1971. Pitfalls in oil spill characterization: Needs for multiple parameter approach and direct comparison with specific parent oils. Presented at Southwest Regional ACS Meeting, San Antonio, Texas, December.

Korringa, P. 1973. The ocean as final recipient of the end products of the continent's metabolism. Pollution of the oceans: Situation, consequences, and outlooks to the future, pp. 91–140. *In* Okologie und Lebensschutz in internationaler Sicht. Verlag Rombach, Freiburg.

Kreider, R. E. 1971. Identification of oil leaks and spills, pp. 119–124. *In* Proceedings, Joint Conftrence on Prevention and Control of Oil Spills. American Petroleum Institute, Washington, D.C.

La Roche, G. 1973. Toxic responses in aquatic organisms. *In*

N. I. Sax, ed. Dangerous Properties of Industrial Material, 4th ed. Van Nostrand Reinhold Co., New York.

La Roche, G., R. Eisler, and C. M. Tarzwell. 1970. Bioassay procedures for oil and oil dispersant toxicity evaluation. J. Water Pollut. Control Fed. 42(11):1982–1989.

La Roche, G., G. R. Gardner, R. Eisler, E. H. Jackim, P. P. Yevich, and G. E. Zaroogian. 1973. Analysis of toxic responses in marine poikilotherms, pp. 199–216. In G. Glass, ed. Bioassay Techniques and Environmental Chemistry. Ann Arbor Science Publishers, Ann Arbor, Mich.

Lee, R. F., and A. A. Benson. 1973. Fate of petroleum in the sea-biological aspects. Background paper, this workshop.

Lee, R. F., R. Sauerheber, and A. A. Benson. 1972a. Petroleum hydrocarbons: Uptake and discharge by the marine mussel, *Mytilus edulis.* Science 177:344–346.

Lee, R. F., R. Sauerheber, and G. H. Dobbs. 1972b. Uptake, metabolism and discharge of polycyclic aromatic hydrocarbons by marine fish. Mar. Biol. 17:201–208.

Levy, E. M. 1970. A shipboard method for the estimation of bunker C in sea water. Paper presented at Symposium on Marine Sciences. Chemical Institute of Canada, Charlottetown, P.E.I., August 16–18.

Levy, E. M. 1971. The presence of petroleum residues off the east coast of Nova Scotia, in the Gulf of St. Lawrence and the St. Lawrence River. Water Res. 5:723–733.

Levy, E. M. 1972. Evidence for the recovery of the waters off the east coast of Nova Scotia from the effects of a major oil spill. Water Air Soil Pollut. 1:144–148.

Levy, E. M., and A. Walton. 1973. Dispersed and particulate petroleum residues in the Gulf of St. Lawrence. J. Fish. Res. Board Can. 30:261–267.

Lie, U. 1968. A quantitative study of benthic infauna in Puget Sound, Washington, U.S.A. in 1963–1964. Fisk Dir. Skr. (Ser. Hauunders) 14:229–256.

Lukens, H. R., D. E. Bryan, N. A. Hiatt, and H. L. Schlesinger. 1971. Development of nuclear analytical techniques for oil-slick identification, Phase IIA. Prepared for U.S. Atomic Energy Commission, June.

Mackie, P. R., A. S. McGill, and R. Hardy. 1972. Diesel oil contamination of brown trout (*Salmo trutta* 1.). Environ. Pollut. 3:9–16.

Martin, R. L. 1966. Fast and sensitive method for determination of nitrogen. Anal. Chem. 38:1209–1213.

McAuliffe, C. D. 1969. Solubility in water of normal C_9 and C_{10} alkane hydrocarbons. Science 158:478–479.

McAuliffe, C. D. 1971. GC determination of solutes by multiple phase equilibration. Chem. Technol. 1:46–51.

McKinney, C. R., J. M. McCrea, S. Epstein, H. W. Allen, H. C. Ureg. 1950. Improvements in mass spectrometers for measurement of small differences in isotope abundance ratios. Rev. Sci. Instrum. 21:724–730.

Miller, J. W. 1973. A multiparameter oil pollution source identification system, pp. 195–203. In Proceedings, Joint Conference on Prevention and Control of Oil Spills. American Petroleum Institute, Washington, D.C.

Mironov, O. G., and L. A. Lanskaya. 1969. Growth of marine microscopic algae in seawater contaminated with hydrocarbons. Biol. Morya 17:31–38. (In Russian)

Mommessin, P. R., A. Hood, W. E. Ellington, and A. F. Mannel. 1968. Initial reports of the deep sea drilling project, Vol. 1, pp. 468–476. Government Printing Office, Washington, D.C.

Nichols, F. H. 1973. A review of benthic faunal surveys in San Francisco Bay. Geol. Surv. Circ. 677.

Nuzzi, R. 1973. Effects of water soluble extracts of oil on phytoplankton, pp. 809–714. In Proceedings, Conference on Prevention and Control of Oil Spills. American Petroleum Institute, Washington, D.C.

Palek, J., W. A. Burby, and F. J. Lionetti. 1972. Size dependence of ghosts from stored erythrocytes on calcium and adenosinel triphosphate. Blood 40:261–275.

Patrick, R. 1949. A proposed biological measure of stream conditions based on a survey of the Conestoga Basin. Lancaster County, Pa. Proc. Acad. Natl. Sci. Phila. 101:277–341.

Ramsdale, S. J., and R. E. Wilkinson. 1968. Identification of petroleum sources of beach pollution by gas liquid chromatography. J. Inst. Pet. 54:326.

Rasmussen, R. A., and F. W. Went. 1965. Volatile organic material of plant origin in the atmosphere. Proc. Natl. Acad. Sci. 53:215–220.

Reish, D. J. 1959. A discussion of the importance of the screen size in washing quantitative marine bottom samples. Ecology 40:307–309.

Robertson, B., S. Arhelger, R. A. T. Law, and D. K. Button. 1973. Hydrocarbon biodegradation in Port Valdez. In Port Valdez Environmental Studies Institute of Marine Science Report, Fairbanks.

Robinson, C. J. 1971. Low resolution mass spectrometric determination of aromatics and saturates in petroleum fractions. Anal. Chem. 43:1425–1434.

Sanders, H. L. 1960. Benthic studies in Buzzards Bay. II. The structure of the soft-bottom community. Limnol. Oceanogr. 5:138–153.

Sanders, H. L., J. F. Grassle, and G. R. Hampson. 1972. The West Falmouth oil spill. I. Biology. Woods Hole Oceanographic Institution Tech. Rep. 72–20. Woods Hole, Mass.

Schink, D. R., N. L. Guinasso, Jr., S. S. Sigalove, and N. E. Cima. 1971. Hydrocarbons under the Sea—A New Survey—Techniques. Offshore Technology, Houston, Texas, April 19–21, 1971. Paper No. OTC1339 1:130–142.

Shannon, C. E., and W. Weiner. 1963. Mathematical Theory of Communications. University of Illinois Press, Urbana. 117 pp.

Simard, R. G., I. Hasegawa, W. Bandaruk, and C. E. Headington. 1951. Infrared spectrophotometric determination of oils and phenols in water. Anal. Chem. 23:1384.

Smith, H. M. 1968. Qualitative and quantitative aspects of crude oil composition. Bull. 642. Bureau of Mines, U.S. Department of the Interior.

Smith, J. B. 1972. Gulf Coast United States crude oil and bottom sediment analysis. Report 721–28–1–72, August.

Sokal, R. R., and P. H. A. Sneath. 1963. Principles of Numerical Taxonomy. W. H. Freeman, San Francisco. 359 pp.

Southwood, T. R. E. 1966. Ecological Methods. Methuen and Company, London.

Soviet. 1973. The weight method of measuring the content of oil products in the sea. February 12, 1973. IMCO paper No. MP XV/3/2/. Intergovernmental Maritime Consultative Organization, London.

Stephenson, W., W. T. Williams, and S. D. Cook. 1971. Computer analyses of Petersen's original data on bottom communities. Ecol. Monogr. 42:387–415.

Strand, J. A., W. L. Templeton, J. A. Lichatowich, and C. W. Apts. 1971. Development of toxicity test procedures for marine phytoplankton, pp. 279–286. Proceedings, Joint Conference on Prevention and Control of Oil Spills. American Petroleum Institute, Washington, D.C.

Straughan, D., ed. 1971. Biological and oceanographical survey of the Santa Barbara Channel oil spill, 1969–1970. Volume I. Biology and Bacteriology. Alan Hancock Foundation, University of Southern California, Los Angeles. 426 pp.

Strickland, J. D. H., and T. R. Parsons. 1965. A practical handbook of seawater analysis. Bull. No. 167. Fish. Res. Board Can., Ottawa. 311 pp.

Swinnerton, J. W., and Linnenbom. 1967. Gaseous hydrocarbons in sea water: Determination. Science 156:1119–1120.

Thode, H. G., J. Monster, and H. B. Dunford. 1958. Sulpher isotope abundance in petroleum and associated materials. Am. Assoc. Pet. Geol. Bull. 42:2619–2641.

Thomas, M. L. H. 1973. Effects of Bunker C oil on intertidal and lagoonal biota in Chedabucto Bay, Nova Scotia. J. Fish. Res. Board Can. 30:83–90.

Thurston, A. D., and R. W. Knight. 1971. Characterization of crude and residual-type oils by fluorescence spectroscopy. Environ. Sci. Technol. 5:64.

Tissier, M., and J. I. Oudin. 1973. Characteristics of naturally occurring and pollutant hydrocarbons in marine sediments, pp. 205–214. *In* Proceedings, Joint Conference on Prevention and Control of Oil Spills. American Petroleum Institute, Washington, D.C.

Voyer, R. A., and G. E. Morrison. 1971. Factors affecting respiration rates of winter flounder (*Pseudopleuronectes americanus*). J. Fish. Res. Board Can. 28:1907–1911.

Wihlm, J. L., and T. C. Dorris. 1966. Species diversity of benthic macroinvertebrates in a stream receiving domestic and oil refining effluent. Am. Midl. Nat. 76:427–449.

Williams, W. T. 1971. Principles of clustering. Annu. Rev. Ecol. Sys. 2:303–326.

Youngblood, W. W., and M. Blumer. 1973. Alkanes and alkenes in marine benthic algae. Mar. Biol. 21:163–172.

Youngblood, W. W., M. Blumer, R. L. Guilard, and R. Fiore. 1971. Saturated and unsaturated hydrocarbons in marine benthic algae. Mar. Biol. 8(3):130–201.

Zafiriou, O. C., J. Myers, R. Bourbonniere, and F. J. Freestone. 1973. Oil spill-source correlation by gas chromatography: An experimental evaluation of system performance, pp. 153–159. *In* Proceedings, Joint Conference on Prevention and Control of Oil Spills. American Petroleum Institute, Washington, D.C.

Zitko, V., and W. V. Carson. 1970. The characterization of petroleum oils and their determination in the aquatic environment. Tech. Rep. 217. Fish. Res. Board Can. Biol. Stn., St. Andrews, New Brunswick.

ZoBell, C. E. 1969. Microbial modification of crude oil in the sea, pp. 317–326. *In* Proceedings, Joint Conference on Prevention and Control of Oil Spills. American Petroleum Institute, Washington, D.C.

ZoBell, C. E. 1971. Sources and biodegradation of carcinogenic hydrocarbons, pp. 441–451. *In* Proceedings, Joint Conference on Prevention and Control of Oil Spills. American Petroleum Institute, Washington, D.C.

3 Fates

Petroleum introduced to the marine environment from whatever sources goes through a variety of transformations involving physical, chemical, and biological processes. This section attempts to identify the major factors controlling each of these processes, to review the relevant experimental and field evidence for quantitative evaluation of the effect of these various degradation processes on petroleum, and to estimate the amount of petroleum hydrocarbons in the sea at present.

Physical and chemical processes begin to operate immediately when petroleum is spilled at the ocean surface. These include evaporation, spreading, emulsification, solution, sea-air interchange, and sedimentation. Involved in all these "physical" processes are chemical factors determined by the composition of the petroleum fraction. In addition, purely chemical oxidation of some of the components of petroleum can be induced by sunlight. The products of these processes include hydrocarbon fractions introduced to the atmosphere, slicks and tar lumps on the surface of the ocean, dissolved and particulate hydrocarbon materials in the water column, and adsorbed or particulate hydrocarbon materials in the sediments.

At the same time physical and chemical separation and degradation processes are occurring, biological processes also act on these various fractions of the original petroleum in various ways. The biological processes considered include degradation by microorganisms to carbon dioxide or organic material in intermediate oxidation stages, uptake by larger organisms and subsequent metabolism, storage, or discharge. In addition, the question of food-web magnification (as has been established for terrestrial fates of chlorinated hydrocarbon residues) is addressed.

Physical and Chemical

Physical and Chemical Characteristics of Petroleum

Crude oils and refined petroleum show an enormous complexity and variation in detailed chemical composition. Since the physical and chemical processes that such petroleum products undergo in the marine environment depend on this composition, a knowledge of at least the major component types and their properties is an essential prerequisite for predicting the fate of petroleum in a more general sense.

The types of petroleum or petroleum products most

likely to be released into the marine environment are crude oils, Bunker C or No. 6 fuel oils, diesel or No. 2 fuel oils, and light petroleum products such as kerosenes or gasolines.

Crude oils of different geologic and geographic sources vary widely in composition. Thousands of individual compounds, mostly hydrocarbons, are found in each crude, and varying proportions of these compounds determine the physical as well as the chemical properties of the oils. Using gross compositional data on all world crudes, the following approximate composition may be determined for the "average" crude oil:

By molecular size
gasoline (C_5–C_{10}), 30 percent; kerosene (C_{10}–C_{12}), 10 percent; light distillate oil (C_{12}–C_{20}), 15 percent; heavy distillate oil (C_{20}–C_{40}), 25 percent; residuum oil ($>C_{40}$), 20 percent

By molecular type
paraffin hydrocarbons (alkanes), 30 percent; naphthene hydrocarbons (cycloalkanes), 50 percent; aromatic hydrocarbons, 15 percent; nitrogen, sulfur, and oxygen-containing compounds (NSOs), 5 percent

Crudes from different sources, however, can differ appreciably from these average values. For example, an average Venezuelan crude would contain about 10 percent gasoline, 5 percent kerosene, 20 percent light distillate oil, 30 percent heavy distillate oil, and 35 percent residuum. Alternatively, the same Venezuelan crude might contain 10 percent paraffins, 45 percent naphthenes, 25 percent aromatics, and 20 percent NSOs. In contrast, south Texas crudes are shifted to smaller molecular size and greater amounts of paraffin–naphthene hydrocarbons than the worldwide crude average. Bunker C or No. 6 fuel oils are the heaviest distillate fractions of petroleum, with specific gravities near 1.00, or greater than viscosity of about 1,000 centipoises (38 °C), and pour point (congealing temperature) around 21 °C. Chemically, the great majority of the compounds are in the C_{30}^+ range, typically 15 percent paraffins, 45 percent naphthenes, 25 percent aromatics, and 15 percent polar NSOs.

Diesel or No. 2 fuel oils represent a middle distillate fraction of petroleum, mixtures of virgin and catalytically or thermally cracked components. Specific gravities usually range from 0.825 to 0.850, viscosity about 40 centipoises, and pour point about −20.5 °C. Chemically, these oils are comprised almost entirely of hydrocarbons in the carbon-number range C_{12}–C_{25}, with greatest abundance at about C_{15}–C_{16}. By molecular type, 30 percent are paraffins, 45 percent naphthenes, and 25 percent aromatics. Some unusual diesel oils may contain as high as 40 percent aromatics.

Light petroleum products, such as kerosene or gasoline, are also made up of virgin and cracked components. Kerosene contains hydrocarbons in the C_{10}–C_{12} molecular weight range, its average specific gravity is about 0.800, and its viscosity is about 1–2 centipoises. A typical kerosene is 35 percent paraffins, 50 percent naphthenes, and 15 percent aromatics. Gasoline contains hydrocarbons in the C_5–C_{10} molecular weight range, its average specific gravity is about 0.700, its viscosity less than 1 centipoise. Virgin gasolines contain about 50 percent paraffins, 40 percent naphthenes, and 10 percent aromatics. Blended gasolines obtained by combining virgin gasolines with catalytically cracked and thermally reformed gasolines contain higher amounts of aromatics (20–30 percent).

Processes

SPREADING

When petroleum or petroleum products are released on the ocean's surface, weathering processes immediately begin to alter the material that spreads out over the surface of the ocean. The extent of spreading is affected by wind, waves, and currents, but probably more by the physical and chemical nature of oil.

To properly assess the behavior of petroleum spills at the air–sea interface, its area of coverage, thickness, and physical condition must be determined as a function of time. Fay (1969), considering the spread of oil on a calm sea, concluded that gravitational effects controlled spreading in the initial stages of a spill. Other factors dictate spreading characteristics as the oil layer thins. For example, consider a crude oil with an interfacial tension of 30 dyn/cm and a density of 0.05 g/cm^3. If such an oil is spread on water, surface forces become influential once the oil has thinned to 0.8 cm.

In field experiments, on the other hand, the area occupied by an oil slick is observed to increase rapidly under the influence of both hydrostatic and surface forces. This area, which depends empirically on the volume spilled (Figure 3-1), eventually becomes constant as further enlargement is limited and offset by natural dispersive forces and other mechanisms. From available spill data, it is apparent that even the viscous crudes spread rapidly into thin layers that are then influenced by surface effects. For example, Berridge *et al.* (1968a,b) using an equation derived by Blokker (1964) (Table 3-1), calculate that 100 m^3 of various crudes will thin to an average value of 0.055 cm after only 17 min, 0.012 cm after 3 h, and 0.003 after 28 h.

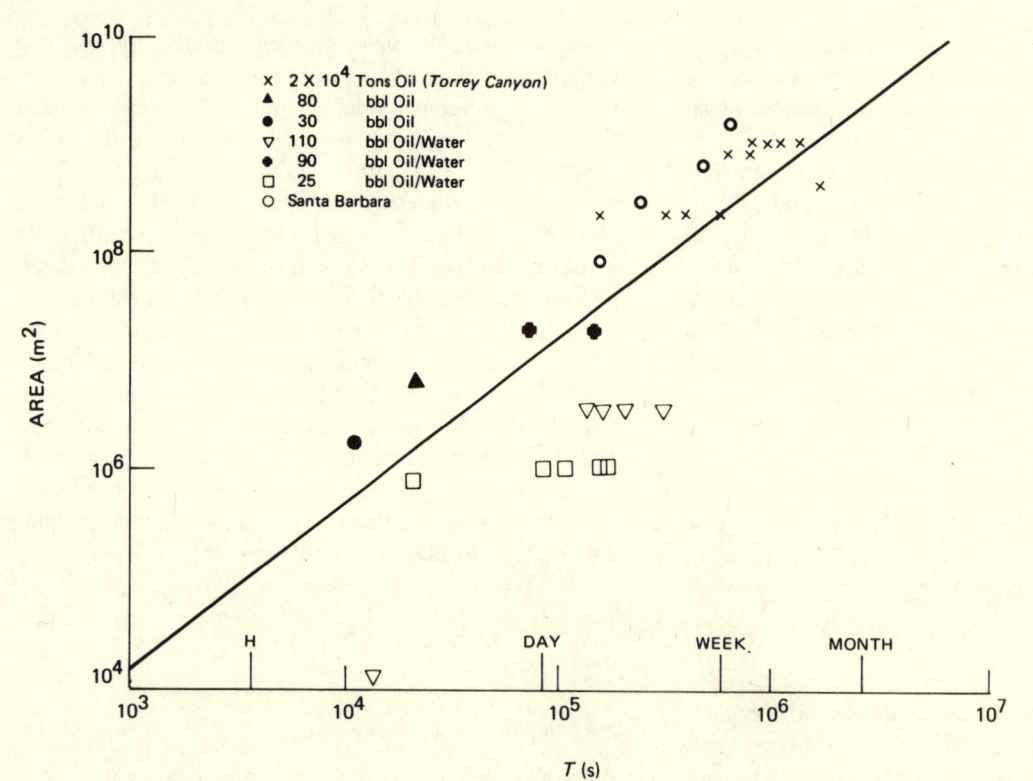

FIGURE 3-1 Increase in slick area due to surface-tension spreading for various-sized spills (Hoult, 1972).

Allen (1969) estimates a minimum average thickness of 0.0025 cm for the Santa Barbara incident, assuming that spreading had occurred for 24 h. Similarly, Smith (1968) estimated that the massive 10 × 40 mi oil slick observed 6 days after the *Torrey Canyon* spill would have had an average thickness of 0.003 cm, a value strikingly similar to other above-cited estimates. Thus even the large spills eventually spread into thin layers. It should be noted, however, that these thickness estimates are averages. Once a spill has thinned to the point that surfaces forces begin to play an important role, the oil layer is no longer continuous and uniform but becomes fragmented by wind and waves into islands and windrows where thicker layers of oil are in equilibrium with thinner films rich in surface active compounds. Observations (Figure 3-2) of experimental spills (P. G. Jeffrey, 1973; Hollinger and Mennella, 1973) have shown that with time one or more patches of thick oil (several millimeters thick) were surrounded by a much larger area of thin film (less than 4 μm). Approximately 90 percent of the oil volume was located in these thicker layers that occupied only 10 percent of the visible slicked area of the sea (Figure 3-3).

In addition, surface currents driven by wind, waves, and convectional cells determine the shape and direction of movement of the spill, wind being the most influential external factor (Blokker, 1964). The oil drift velocity is about 3 percent that of the wind speed. According to Hoult (1972) the center of mass of the oil moves according to $d\vec{x}/dt = \vec{U}_{\text{current}} + 0.035\vec{U}_{\text{wind}}$ where $\vec{x}$ is the coordinate vector of the center of mass and $\vec{U}$ velocity. The drift due to wind and wind-driven waves is accounted for by the latter term in the expression.

The importance of the surface active constituents in spreading and dispersion of oil on water is demonstrated by the fact that most pure hydrocarbons do not spread spontaneously by surface forces. Only aromatic and aliphatic hydrocarbons more volatile than *n*-nonane have positive spreading coefficients, while none of the cyclic hydrocarbons will spread by surface forces. The nitrogen–sulfur–oxygen (NSO) compounds in petroleum and petroleum products are likely to play an important role in the fate of oil released in the marine environment. Many of these compounds are highly surface active, such as the cyclic carboxylic acids identified by Seifert and Howells (1969) in California crudes. These surface active constituents are immediately important in their effect on the spreading characteristics and emulsifiability of petroleum products. Thus, they enhance the weathering process by increasing the surface volume ratio of the oil, thereby producing a greater exposure to air and underlying water.

The cessation of slick growth has also been attributed to a decrease in the spreading tendency of the oil

TABLE 3-1 Blokker Constants Calculated from Observed Maximum Slick Dimension, 120-Ton Spill[a]

Time from $t = 0$ (min)	Maximum Slick Dimension (km)	Blokker Constant (K_r)[b]
0	0.225	
4	0.279	319
17	0.344	199
22	0.366	195
27	0.356	144
33	0.417	194
42	0.505	313
48	0.521	301
98	0.582	189
119	0.654	251
124	0.689	285
422	0.751	109
427	0.876	170
514	0.876	142
517	0.835	125
2820	1.44	117
3088	2.16	360
5760	2.40	265

[a] From P. G. Jeffrey, 1973.
[b] $K_r t = \pi(r^3_t - r^3_0) d_w / 3V(d_w - d_o)d_o$, where K_r is a constant referred to as the Blokker constant, r_t and r_0 are the radii of the slick, assumed to be circular in shape, t is the spreading time in seconds, V is the volume of oil in the slick in cm^3, and d_w and d_o are the density of seawater and the oil.

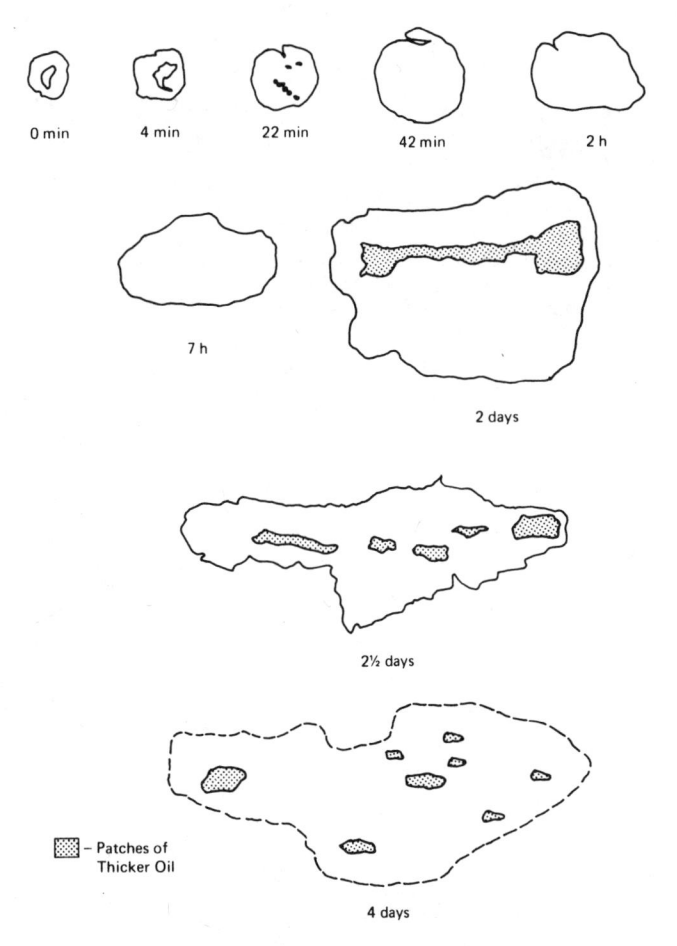

FIGURE 3-2 A series of diagrams showing the outline development and subsequent breakup of the oil slick (P. G. Jeffrey, 1973).

caused by loss of the surface-active NSO compounds by dissolution (Fay, 1971). However, most of these compounds are not much more soluble than hydrocarbons (see below), and it is more likely that weathering of petroleum products containing heavy ends results in an eventual semisolid mass (the precursor to a tar ball) that is no longer fluid and does not spread. This could also be a water-in-oil emulsion (see below). Because the NSO constituents are encased within the semisolid matrix, their loss would be extremely slow. This contention is supported by surface film pressure data (W. D. Garrett, personal communication), that indicate that the concentrations of polar surface active compounds are higher in tar balls than in the parent oils from which they were formed.

EVAPORATION

The greater the rate and extent of spreading, the greater is the rate of evaporation (Hoult, 1972). Simulated evaporation experiments in the laboratory (Kreider, 1971) indicate that all hydrocarbons containing less than about C_{15} (boiling point <250 °C) will be volatilized from the ocean's surface within 10 days (Figure 3-4). Hydrocarbons in the C_{15}–C_{25} range (boiling point 250–400 °C) show limited volatility and will be retained for the most part in the oil slick. Above C_{25} (boiling point >400 °C) there will be very little loss.

Thus, evaporation alone will remove about 50 percent of the hydrocarbons in an "average" crude oil on the ocean's surface. A Bunker C or No. 6 fuel oil will probably lose less than 10 percent. Indeed, as evaporation proceeds, a crude oil begins to resemble a Bunker C oil in carbon-number distribution. Evaporation will greatly affect the diesel or No. 2 fuel oil spill, with 75 percent or more being rapidly volatilized. Only traces of the lightest petroleum products (kerosene, gasoline) will be retained in the water.

If the petroleum release occurs in the open sea, evaporation may well be virtually complete before the material reaches shorelines. Rough seas would tend to increase evaporation rates, since sea spray and bursting bubbles would eject both volatile and nonvolatile components together into the marine atmosphere.

These evaporated hydrocarbons enter the atmospheric pool of hydrocarbons, and very little is likely to return

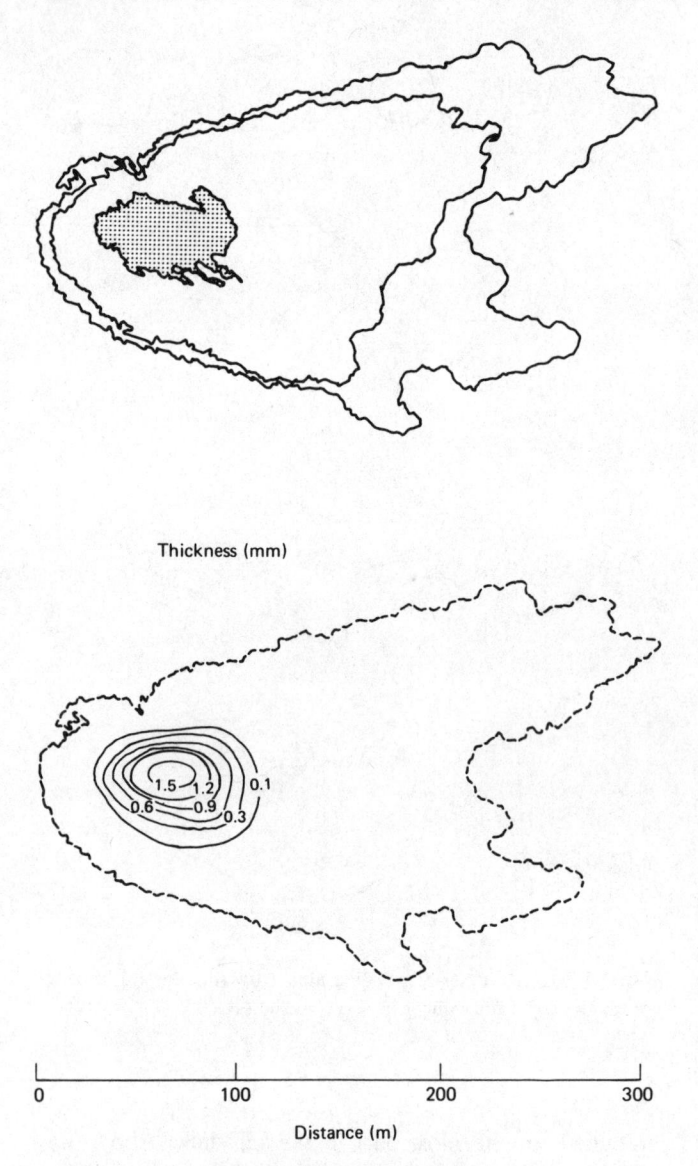

FIGURE 3-3 Comparison of visible slick with actual thickness of oil on water as measured by multifrequency microwave radiometry (Hollinger and Mennella, 1973).

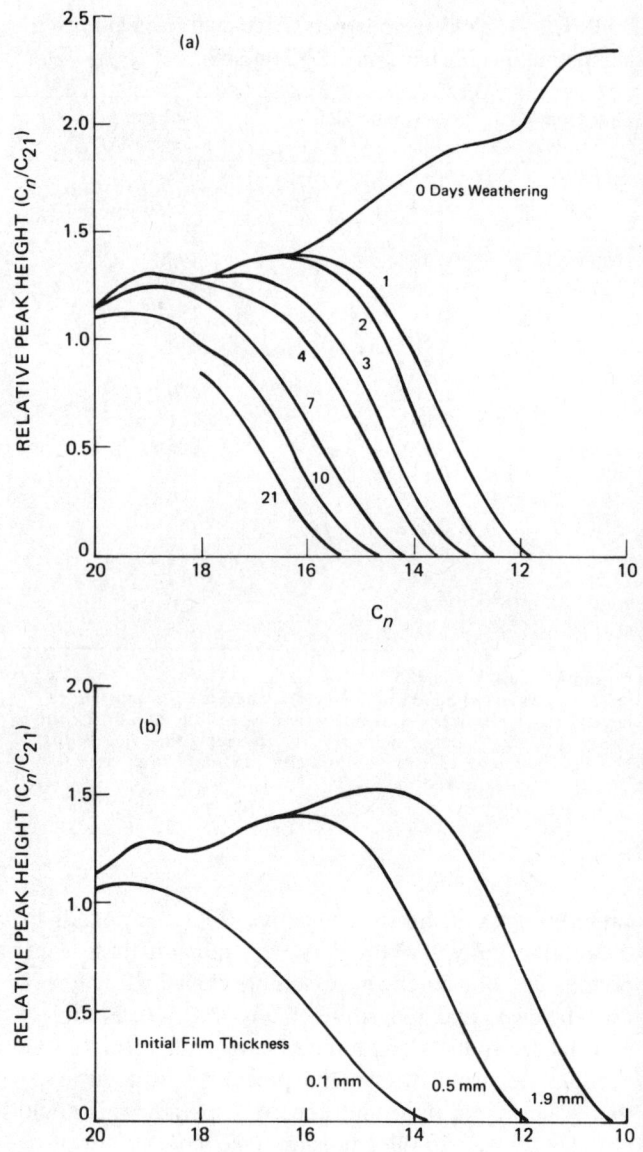

FIGURE 3-4 Simulated weathering of crude oil (Kreider, 1971). (a) Effect of time on weathering, 0.5 mm initial film thickness. (b) Effect of film thickness on 24-h weathering.

to the oceans as hydrocarbons. Chemical reactions in the atmosphere, such as photocatalytic oxidations, convert an unknown amount of these hydrocarbons into less volatile nonhydrocarbon compounds that may reenter the oceans. The fate and effect of these types of compounds are unknown.

SOLUTION

Solution is another physical process in which the low-molecular-weight hydrocarbons, as well as some of the more polar nonhydrocarbon compounds are lost from the surface to the water column. The rate of this process is governed by the wind and sea state and petroleum material (chemical composition, gravity, viscosity, pour point, surface tension, etc.). Although this solution process starts immediately, it has long-term aspects as well. For example, biological and chemical oxidation processes produce polar compounds (e.g., alcohols, fatty acids) from hydrocarbons in the oil.

Solution and evaporation both affect the lower molecular weight range hydrocarbons. However, the water solubility (McAuliffe, 1966, 1969a; Baker, 1967;

Davis et al., 1942) of hydrocarbons drops drastically as one goes to higher carbon numbers. For example, of the normal paraffin hydrocarbons, n-C_5 has a distilled water solubility of about 40 ppm; n-C_6, about 10 ppm; n-C_7, about 3 ppm; n-C_8, about 1 ppm; n-C_{12}, about 0.01 ppm; and n-C_{30}, about 0.002 ppm. For aromatic hydrocarbons, C_6 (benzene) has a distilled water solubility of about 1,800 ppm; C_7 (toluene), about 500 ppm; C_8 (xylenes), about 175 ppm; C_9 (alkylbenzenes), about 50 ppm; C_{14} (anthracene), about 0.075 ppm; and C_{18} (chrysene), about 0.002 ppm. Hydrocarbon solubilities in seawater are probably not more than 12–30 percent smaller than those in distilled water (Harned and Owen, 1958).

Thus, the most important effect of solution is to remove the light saturated and aromatic hydrocarbons (up to about C_{10}) rather rapidly. Simultaneously, of course, the seawater will gradually leach out higher molecular weight hydrocarbons. From the solubility data above it does not appear that the higher molecular weight aromatics (C_{14}^+) are appreciably more soluble than the paraffins of comparable molecular weight. Therefore, these aromatic hydrocarbons in the medium to high molecular weight may not be preferentially extracted by water leaching.

As mentioned previously, hydrocarbon degradative processes, either chemical or microbiological, can produce polar compounds with appreciably greater water solubility than the original hydrocarbons had. For example, naphthalene (solubility, 32 ppm) can be oxidized to α-naphtol (solubility, 740 ppm; Seidell, 1941). Thus, as these oxidation products are formed, they may be leached out of the petroleum material more rapidly. Many if not most of the high-molecular-weight polar NSO compounds naturally occurring in petroleum, although surface active, are *not* appreciably more water soluble than are hydrocarbons of comparable molecular weight. Structural studies that have been made on these compounds show that they are really *hydrocarbonlike, except for an isolated heteroatom* (nitrogen, sulfur, oxygen), usually only one and rarely more than two, in the molecule (Johnson and Aczel, 1967).

EMULSIFICATION

Because most of the components of petroleum are relatively insoluble in water, an important mode of dispersion is the formation of emulsions. Oil-in-water emulsions are dispersed easily by currents and turbulence at the surface, particularly in rough seas. These fine particles of oil have been investigated in detail by Forrester (1971) following the *Arrow* spill in Chedabucto Bay, Nova Scotia. Particles ranging from 5 μm to several millimeters were found in the water column and were distributed as far as 250 km from their source.

Water-in-oil emulsions are also formed, particularly from heavy asphaltic crudes or residual oils. These tend to be more coherent semisolid lumps, referred to as "chocolate mousse" in writings on the *Torrey Canyon* spill. Experimental and limited field studies (Berridge et al., 1968a,b) have shown that these persistent emulsions contain approximately 80 percent water and that bacteria or solid particulate matter do not seem to be required for their formation. Naturally occurring surfactants in the petroleum (as discussed above) or dispersants added to control spilled oil will, of course, tend to assist in the formation of emulsions of both kinds.

More recent field studies (Dodd, 1971) showed that the rate of formation of water-in-oil emulsions under comparable conditions varied dramatically with the nature of the oil: Kuwait crude took up 40 percent water in 2–4 h, whereas Tia Juana crude took 2 days to reach 3 percent water and 8 days to reach 57 percent.

The eventual fate of oil-in-water emulsions would appear to be dissolution in the water column or association with solid particulate matter or detritus and eventual biodegradation or incorporation in sediments; the water-in-oil "mousse" has been suggested (Berridge et al., 1968a) as a source of beach tar and presumably also of pelagic tar lumps.

DIRECT SEA–AIR EXCHANGE

Petroleum hydrocarbons are removed from the sea surface not only by evaporation but also by wave-produced spray and bursting bubbles. Quantitative estimates of global removal by these mechanisms are virtually impossible to make since the transfer into the atmosphere will depend on wind speed, sea state, and the extent to which wave breaking and whitecap formation are suppressed by oil films. Other considerations include the coagulation of bubbles before bursting in the oil film, and modification, by the oil, of the number and size distribution of jet and film particles produced by the bursting bubbles. There is some question as to whether whitecap suppression by oil films is due to the reduction of wave breaking or to the more rapid bursting of bubbles through oil films, which would make the whitecap less visible. The suppression of wave breaking and whitecaps will depend largely on film thickness and horizontal extent of the film.

Baier (1970) reported that oil films, accumulated on the surface of a freshwater lake after recreational boat use, disappeared within 3 days. This may have been simply evaporation, since these oils are mostly lighter than C_{15}. However, on the basis of laboratory studies,

Baier (1972) suggested that the removal of "thin" oil films of this type can be due largely to bubble bursting and spray formation. This process seems to be accelerated by solar radiation, which can cause the formation of polar, surface active molecules in the film.

Sea–air transfer processes of this type are most effective for the removal of relatively thin films. Eriksson (1959) has calculated that particles with radii less than 50 μm are injected into the atmosphere by bubble bursting and spray: approximately 3×10^{-10} g of seawater/cm^2 s for a 6 m/s wind speed.

MacIntyre (1970) has shown that the particles produced by bursting bubbles are formed on the liquid surface. If this thickness applies to particles produced from water surfaces covered by a thin oil film as well and if an oil film of 1μm thickness does not suppress the formation of whitecaps, such a film could be removed from the water surface by this mechanism in approximately 3–4 days at a wind speed of 6 m/s. Removal would be more rapid at higher wind speeds, while thicker oil films would take longer to disperse.

Garrett (1968) and Paterson and Spillane (1969) have shown that the presence of insoluble monomolecular films will enhance the production of atmospheric particles when clusters of bubbles burst at a water surface, but it is not clear whether this would increase the removal rate of film material at the air–sea interface. Of course, this method of dispersal is only important for the removal of very thin oil films, which may be expected primarily at the periphery of oil spills.

It should be emphasized that, except in coastal areas with onshore winds, removal of hydrocarbons from the ocean surface by this mechanism is temporary. The particles injected into the atmosphere have relatively short residence times, ranging from seconds for particles greater than 100–200 μm radius up to a few days for particles with radii of a few micrometers. Thus, most particles will be redeposited in the ocean at distances ranging from a few meters to several hundred kilometers from their point of injection into the atmosphere.

Hydrocarbons in nonspill areas are also injected into the atmosphere by bubble bursting. Hydrocarbons are not surface active themselves, but may tend to be solubilized by naturally occurring surfactants such as humic substances, fatty acids, alcohols that are concentrated in the sea surface microlayer. There are very few measurements of hydrocarbons in the surface microlayer. Hydrocarbon enrichment in the top 150–200 μm of the water surface compared with water 20 cm below the surface has been reported (Duce et al., 1972) in an estuarine area. In open ocean regions of the North Atlantic, hydrocarbon enrichment up to a factor of 2.5 in the surface microlayer was found in 9 of 15 samples collected in nonspill areas (J. G. Quinn, personal communication). Hydrocarbon enrichment in the surface microlayer in nonspill areas is apparently not a universal feature of open ocean water.

PHOTOCHEMICAL OXIDATION

The rate of oxidation of hydrocarbons and NSO compounds in petroleum varies with their chemical nature (Monaghan and Koons, 1973). For example, alkyl substituted cycloalkanes tend to be oxidized more rapidly than normal paraffins. These "autooxidation" rates may be accelerated by photolysis initiation followed by free radical chain steps, or by catalysis by metallic ions of variable valence and may be slowed by chain stopping steps due to sulfur atoms (Dodd, 1971). In photochemical processes, the optical density of the oil, particularly in the ultraviolet, is an important variable.

Experimental estimates have been made of the actual removal rates of slicks attributable to photolysis (Freegarde and Hatchett, 1970). Using light sources resembling the spectrum of sunlight, within a factor of two in internal relative intensities, the total rate of decomposition corresponds to the destruction of a 2.5 μm slick in 100 h.

A slick of this thickness has about 2,000 kg/km^2 (Garrett, 1969). Assuming an effective 8-h day of sunlight, photolysis can initiate sufficient oxidation reaction to remove the slick in a few days. This is probably a conservative removal estimate since it assumes fairly complete oxidation, whereas formation of carboxylic or phenolic acids increases the solubility and may also increase biodegradation and emulsification rates.

TAR LUMP FORMATION

The occurrence of petroleum residues on beaches and the open sea is well documented (Dennis, 1959; Ludwig and Carter, 1961; Brunnock et al., 1968; Horn et al., 1970; Heyerdahl, 1971; Morris and Butler, 1973; Blumer et al., 1973; Butler et al., 1973; L. M. Jeffrey, 1973; Monaghan and Koons, 1973).

The composition of these tar lumps is quite varied, but most of them include paraffinic hydrocarbons up to C_{40} (Brunnock et al., 1968; Adlard et al., 1972; Morris and Butler, 1973) with either a single maximum typical of a weathered crude oil (80 out of 110 chromatograms) or a double maximum characteristic of a waxy crude oil sludge (30 out of 110) (Butler et al., 1973; Figure 3-5). Very little material lighter than C_{15} is ever found, but 20–40 percent of the material is nonvolatile in 10 min at 200 °C in the injector port of the gas chromatograph (Levy, 1972).

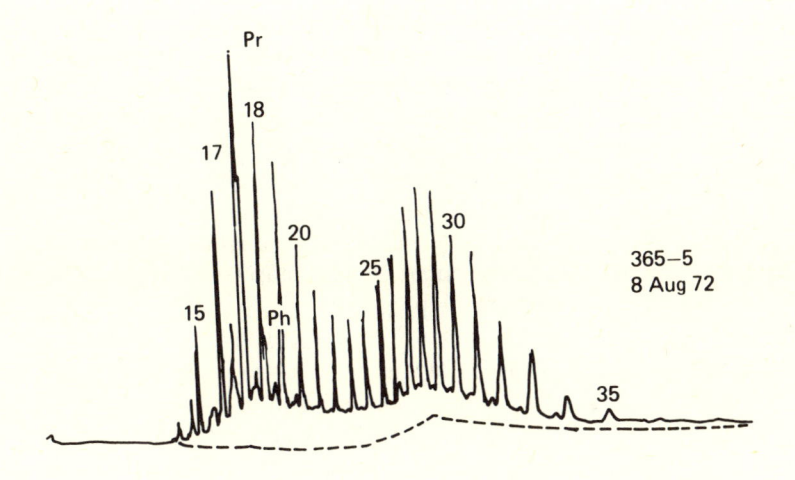

FIGURE 3-5 Gas chromatographs of tar lumps showing high boiling paraffinic wax components as a second maximum in the envelope of peaks (Morris and Butler, 1973).

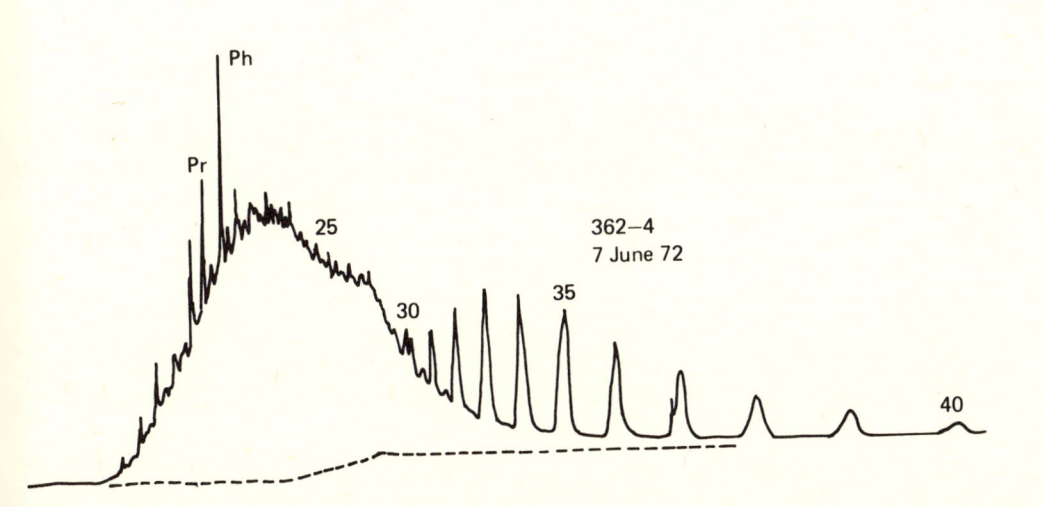

Liquid chromatographic and mass spectrometric studies (Monaghan and Koons, 1973) of tar collected east of Galveston Bay along the Texas coast show high NSO compound contents (average 45 percent, maximum 70 percent). Less specific analyses for resins and asphaltenes (60–66 percent) in Bermuda beach tar have also been reported (Blumer et al., 1973).

Chloroform-insoluble material from tar lumps collected off Florida (Attaway et al., 1973) consists of pulpy organic matter and dense mineral-like particles as well as up to 35 percent water. Some of these mineral particles were insoluble in acid and seemed to be quartz, but the soluble portion was high in iron oxides. The iron oxide content of tar on a dry weight basis ranged to 18.7 percent, with 24 out of 35 samples containing more than 0.1 percent Fe_2O_3. Since the iron content of most crude oils is less than 100 ppm, these results imply that a substantial fraction of pelagic tar lumps contain rust and were the product of discharges from steel tanks or other steel apparatus (e.g., offshore drilling or wastes from engines either on land or in ships).

It seems likely that the frequent occurrence of waxy material resembling crude oil sludge is also an indication that the source of some pelagic tar is the washing from tankers (Brunnock et al., 1968; Blumer et al., 1973; Morris and Butler, 1973; Butler et al., 1973).

The content of NSO compounds is higher than would be expected from simple evaporation and dissolution of a crude and may be evidence for photooxidation during the weathering process. As discussed above, aromatic molecules could be oxidized photochemically or by means of trace metal catalysis. This would produce oxygen-containing compounds that would be reported in the NSO fraction. In support of this, most of the polynuclear aromatic compounds and the light absorbing NSO compounds present in the crude oil, as well as most of the heavy metals (Filby and Shah, 1971), would tend to be retained in the tarry residue left after the initial phase of evaporation and dissolution. Another possibility is not simple photooxidation but photolysis of larger molecules (including NSO compounds) with the production of free radicals and hence polymeric materials. This would tend to increase the amount of heavy asphaltic material over that present in the original crude.

Tar balls found on the beach near California seep areas (Ludwig and Carter, 1961) were found to be porous and may have resulted from agglomeration of smaller lumps in the surf zone. In contrast, most tar lumps collected at sea appear to have been formed as single pieces. Inhomogeneities due to weathering are only at the surface. Inclusions of paraffinic wax are frequent; one lump contained a large crystal of yellow wax consisting entirely of C_{30}–C_{40} paraffins (Blumer et al., 1973; Butler et al., 1973). Thus it seems likely that most pelagic tar lumps result from a single oil sludge lump or slick residue.

SEDIMENTATION

Deposition of most liquid petroleum compounds in subtidal environments requires an increase in density sufficient to sink the material. Processes increasing density include (a) evaporation and dissolution of the lighter compounds, (b) formation and agglutination of dispersed particles, followed by uptake onto particulate sediment, and (c) absorption or adsorption of dissolved species onto particulate matter.

Evaporation and dissolution, combined with other reactions such as oxidation, lead to the formation of semisolid globules (tar balls). The nature and occurrence of these is discussed above, but Morris and Butler (1973) have suggested that further degradation of the tar surface leads to the formation of small, dense particles that sink; recovery of such particles in the upper several hundred meters confirms a downward dispersion of these materials (B. F. Morris, personal communication).

Formation and agglutination of dispersed liquid petroleum particles have been observed to occur following spills (Forrester, 1971). Particles are dispersed in the water column and come into contact with suspended inorganic sediment. Thus, process (b) can be important in removing dispersed petroleum from surface and nearsurface waters in coastal areas, especially estuaries, which are characterized by a high concentration of suspended sediment (Lisitzin, 1972; Spooner, 1970). A variant of this process occurs when increased bacterial and invertebrate biomass associated with a stabilized slick causes a portion of the slick to sink (Voroshilova and Dianova, 1950).

The sedimentation of petroleum spread on the sea may be substantially dependent on the availability of dense particulate materials to act as nuclei. The chemical nature of such particulate material depends on the locale, particularly the runoff-borne materials. Organic material, clays, calcite or aragonite, glacial flours, siliceous grains, all will intercalate, coacervate, or adsorb oils in different biogeochemical areas and their subsequent availability for biodegradation and other chemical reactions is thus affected. Rough seas may also increase the chances of the petroleum to be absorbed on or mixed with particulate matter (sands, silts, clays, shell fragments, etc.) stirred up in shallow water. These particles eventually settle to the bottom when the seas become calmer. Many Bunker C oils and some heavy crudes (e.g., Venezuela, California) have specific gravities near 1.000, and thus very little particulate matter is needed with these oils to exceed the specific gravity of seawater (1.025).

Studies of process (c) indicate that fine-grained (<44 μm) clay minerals adsorb or absorb the greatest quantity of dissolved hydrocarbons (Meyers, 1972). Organic material (3–5 percent) in the clay may assist this process: Sorption increases with salinity and decreases with temperature and can remove dissolved petroleum from surface marine waters (Suess, 1968).

Rates of incorporation into subtidal sediments, by processes (a) through (c) and augmented by pelleting of particulate matter by organisms, are not known. Following spills, considerable incorporation can occur within a few weeks (Kolpack et al., 1971; Blumer and Sass, 1972b). No data are yet available for rates of petroleum incorporation in deep ocean sediments, although plutonium appears to be sedimented on particles at 70–400 m/year (Noshkin and Bowen, 1973).

The movement of petroleum, once sedimented, is poorly understood. Significant lateral spreading (10 m–1 km) can occur for at least several months after spills (Blumer et al., 1970b, 1971; Blumer and Sass, 1972b; Kolpack et al., 1971). The form of the oil and the mechanisms for dispersal are not known.

Sedimentation in intertidal areas is dependent on substrate type and rates of physical reworking. On beaches, reworking of liquid and particulate petroleum occurs on the foreshore (Ludwig and Carter, 1961; Drapeau, 1970; Asthana and Marlowe, 1970). On high-energy shoreline, the entire amount of tar deposited on one tide cycle can be removed on the next cycle (Butler et al., 1973). Seasonal accretion and erosion cycles allow reworking of buried oil. Increased oxidation and dissolution of petroleum can occur (Guard and Cobet, 1973). Rates of these processes appear to depend on the rate of tidal pumping of seawater through the beach. Tarry globules on beaches tend to accumulate large amounts of sediment, become rounded, and behave finally as pebbles (Ludwig and Carter, 1961).

Also, tar globules tend to concentrate near and above normal high tide. Degradation of large masses is very slow in this latter environment (Blumer *et al.*, 1973).

Low-energy shorelines tend to contain fine-grained sediment, and no seasonal cycles of sedimentation occur (Gebelein, 1971). Sedimentation generally is rapid, so that oil layers or particles can be buried within several months. Mechanical reworking is limited. However, infauna ingest sediments, pelletize materials, and incorporate petroleum into the subsurface. Degradation is limited in the subsurface by anaerobic conditions (Gebelein, 1971; Blumer and Sass, 1972b). Tidal pumping of seawater through the sediments occurs at slow rates relative to beaches (Gardner, 1973).

On rocky shores, deposition of liquid (e.g., from slicks) and finely dispersed petroleum is usually limited to the algal felt in the lower intertidal region and to pores in the rock surface. Tidal flushing is extensive and leads to removal of considerable petroleum over several months (Foster *et al.*, 1971). Petroleum in the algal felt is removed by death of the algae and by browsing (Spooner, 1969). Permanent deposition of particulate petroleum is limited to surfaces of relatively low slope, where materials are stranded and attach to the substrate by partial melting. Removal and degradation of these materials may take several years (Blumer *et al.*, 1973; C. D. Gebelein, personal communication).

Thus, although sedimentation plays an essential part in the fate of oil in the marine environment, virtually no systematic fieldwork has been done on this subject, and it is difficult to make more than rough qualitative predictions about either the rates of sedimentation or the amounts of petroleum to be found in the sediments unless actual field studies have been carried out. A summary table of the limited number of field observations to date is included in a later section.

SUMMARY: THE LIFE HISTORY OF A SPILL

This section represents an attempt to summarize the above discussions in terms of an oil spill on the sea, from its initial fluid, unmodified condition through its final residual forms. As discussed above, the resulting slick is dispersed by physical forces and chemically modified by oxidative and biological processes. As the spill ages the relative impact of the various dispersive and degradative processes shifts from the rapid physical effects to slower chemical and biological modifications. As discussed above, the rate of dispersion of a spill is primarily a function of air–sea dynamics, chemical and physical properties of the oil, and the magnitude of the spill. In spite of the seeming complexity of this problem, it is possible to identify stages in the lifetime of a spill and to assign priorities to the processes acting to modify it. An understanding of anticipated events is essential for designing and deploying oil recovery equipment, interpreting aerial spill surveillance data, and tracing a petroleum residue to its source in order to determine the fate of oil on the sea.

These processes can be divided into three stages. Initially, the fluid spill spreads rapidly under the influence of gravitational and surface chemical effects. The polar, surface active constituents (containing nitrogen, oxygen, and sulfur) are highly influential in spreading the spill into extremely thin layers that approach monomolecular dimensions at the outer edges of the slick. In addition, the extent and rate of slick growth is determined by wind, waves, and current and by the density and viscosity of the spilled petroleum. The rate of most dispersive processes is greatly enhanced by spreading the oil into thin layers. Thus, surface-to-volume ratio of the spill increases and a greater exposure to air, sea, and sunlight results. In this early stage, evaporation is the predominant dispersal process, and its rate increases with winds and the further spreading of the spill. In rough seas with breaking water and wave-induced surface contractions, oil is emulsified into the ocean. The entrained oil may also be sedimented by adsorption onto nonbuoyant particulate matter. Another process occurring in rough weather is the transport of oil into the atmosphere by ejection of oil-coated drops from sea spray and bursting bubbles. Fallout of this oily mist usually returns to the sea downwind and usually is outside of the initial spill. If the spill had occurred in coastal waters, it may fall out on nearby land.

It is difficult to determine whether the dissolution rate will increase or decrease with time. Dissolution is slow for most of the compounds found in petroleum: The more soluble compounds, spread into thin layers initially, will dissolve in the underlying water. Organic acids will be solubilized by reaction with the relatively alkaline seawater, and photocatalyzed oxidation may produce species more volatile or soluble than the parent compound. Processes of this kind achieve a more important dispersant role at a later stage in the life of the spill after the most volatile compounds have been evaporated. In this mesostage, photochemical and biological degradation assumes a more dominant role. The expansion of the slick by spreading has essentially ceased. The rate of dispersion of the thin external portions of the spill is balanced by the rate of spread from thicker central portions. In addition, for crudes and residual fuels, loss of light ends creates a more viscous substance that does not flow readily along the air–water interface.

Thus the density and viscosity of the residue increase,

and eventually the third, most refractory, stage is reached. The size of the resulting tarry residue is large when thick layers have weathered under relatively mild meteorological conditions. Smaller fragments are produced by waves and breaking water. Further degradation, weathering, and interaction with the environment is extremely slow since the surface-to-volume ratio of unspread tar is small, and most dispersive reactions occur at interfaces. More importantly, since the petroleum residue is nonfluid, additional spreading ceases and the internal contents become encapsulated and isolated from effective interaction with dispersive and degradative processes. Microbial degradation becomes important as populations of hydrocarbon-adapted bacteria develop.

Although distillate fuels would undergo many of the dispersive processes listed above, there would be little tendency to form tar balls. However, because most of the petroleum spilled on the sea surface contains nonvolatile, high-molecular-weight components, tar formation is likely. These sources include accidental spills, bilge pumping, and tank-cleaning operations, as well as input from land-based sources and undersea seeps.

It is important to note that although the initial dispersion of a slick is fairly well understood, the longer term process (except possibly for bacterial degradation of normal paraffins) has not been studied under field conditions. Much of what we have presented is thus an extrapolation of processes studied mainly in the laboratory and should be verified by field studies.

Amounts of Hydrocarbons in The Marine Environment

At present, field data are rather sparse, making it difficult to generalize sufficiently to predict the probable levels of petroleum residues in a given area where their effects might be important (e.g., an estuary). In addition, the distinction has not always been made between hydrocarbon materials resulting from oil spills or other anthropogenic input and those resulting from natural sources such as biosynthesis or crude oils seeps. Some criteria for distinguishing anthropogenic (e.g., high iron content, alkyl-aromatic and asphaltic compounds, regular series of homologs) or biogenic (e.g., distinctive paraffinic and olefinic compounds) hydrocarbons are pointed out in Chapter 2 and elsewhere (Attaway et al., 1973; Farrington et al., 1973). Finally, time-series data for petroleum residues are available only for a few locations such as Florida (Dennis, 1959), Bermuda (Butler et al., 1973), and West Falmouth (Blumer et al., 1970b; Blumer and Sass, 1972a); and even in these locations surveys cover only a few years. Nevertheless, this section has been prepared in the interest of stimulating further work. In the following, reservoirs of petroleum (principally hydrocarbon) residues are classed as atmosphere, sea surface microlayer, pelagic tar, water column, and sediments.

ATMOSPHERE

Few of the available measurements of atmospheric hydrocarbon concentration are good measures of the reservoir of petroleum hydrocarbons in the atmosphere over the sea. Furthermore, most of the atmospheric hydrocarbons are low-molecular-weight (methane to C_6) and would not enter other parts of the marine environment without first being oxidized or possibly adsorbed on particulate matter.

For example, Cavanaugh et al. (1969) measured concentrations in the atmosphere at Point Barrow, Alaska, by gas chromatography. Ethane, ethylene, butane, benzene, and other unidentified materials were each less than 1 ppb; the total average was about 6 μg/SCM. R. A. Rasmussen (personal communication) sampled uncontaminated marine air under trade-wind conditions on the northeast side of the island of Hawaii and found 10 μg/m^3 of hydrocarbons in the C_3–C_{12} range. Preliminary measurements (J. G. Quinn and T. L. Wade, personal communication) from a sampling tower on the west coast of Bermuda have shown that approximately 95 percent of C_{14}–C_{33} hydrocarbons are either in the gas phase or are present on particles smaller than about 0.02 μm (two Gelman A glass fiber filters). For several samples collected, the gas phase concentration ranged to 1 μg/SCM (30 percent resolved n-alkanes and branched alkanes; 70 percent unresolved aromatics and naphthenics), and hydrocarbons in the particulate phase were less than 0.1 μg/SCM. An automatic sampler on the R. V. Trident (Duce and Quinn, unpublished manuscript) gave particulate hydrocarbon concentrations ranging from 0.002 to 0.03 μg/SCM in the C_{14}–C_{28} range. All these are very much lower than the typical concentrations of hydrocarbons over land (50–200 μg/SCM).

SEA SURFACE MICROLAYER

Although hydrocarbons themselves are not surface active, they tend to be accumulated at the sea surface by naturally occurring surfactants, as discussed above. We have already described the two known field measurements of hydrocarbons in the surface microlayer in Narragansett Bay (Duce et al., 1972) and in the open North Atlantic (J. G. Quinn and T. L. Wade, personal communication). In both cases, the hydrocarbon enrichment of the surface microlayer over a subsurface sample was not large, at most a factor of 2.5, and in a number of cases no enrichment at all was observed in nonslick areas. In contrast, Keizer and Gordon (1973) report enrichment of the top 1-5 mm by a factor of approximately 20 over samples taken at 1-5 m depth. It would be unwise, however, to generalize from a few preliminary observations, and further studies of this problem are required to firmly establish the role of the microlayer.

One important consideration is that oil slicks or surface films in nonslick areas may act as a differential accumulation center for trace materials (Filby and Shah, 1971; Guinn and Bellanca, 1970) such as metal ions, vitamins, amino acids and such lipophilic pollutants as DDT residues and PCB (Duce et al., 1972; Quinn and Wade, 1972). Whatever ecological effects these materials have on the life near and at the sea surface, they may be substantially altered by the presence of oil slicks; and as a result, the ecology of the sea surface may be changed with possible consequent impacts on other pelagic life (Feldman, 1970a).

PELAGIC TAR LUMPS

In the last few years a number of quantitative measurements of the weight of tar lumps collected by neuston nets have been published or otherwise made available to this panel. The average amounts observed in various portions of the world oceans are summarized in Table 3-2, but no data at all were available for two-thirds of the world oceans. The only time series has been taken at Station S (32°10′ N, 64°30′ W) southeast of Bermuda (Butler et al., 1973), and the variations in biweekly measurements of pelagic tar are shown in Figure 3-6. On two occasions (February and August 1972) replicate tows were run within 24 h

TABLE 3-2 Tar Densities on the World Oceans

Location and Reference	Area (10^{12} m^2)	Tar (mg/m^2) max.	Tar (mg/m^2) mean	Total Tar (10^3 tons)
NW Atlantic Marginal Sea (Morris, 1971; Morris and Butler, 1973)	2	2.4	1	2
East Coast Continental Shelf (Sherman et al., 1973; Attaway et al., 1973)	1	10	0.2	0.2
Caribbean (L. M. Jeffrey, 1973; Sherman et al., 1973)	2	1.2	0.6	1.2
Gulf of Mexico (L. M. Jeffrey, 1973; Sherman et al., 1973)	2	3.5	0.8	1.6
Gulf Stream (Morris, 1971; Morris and Butler, 1973)	8	10	2.2	18
Sargasso Sea (Morris and Butler, 1973; Sherman et al., 1973; Attaway et al., 1973; Polikarpov et al., 1971)	7	40	10	70
Canary and North Equatorial Current (Heyerdahl, 1971)	3	1,000?	?	? (large)
Rest of northeast Atlantic	8	?	?	?
North Sea	3	?	?	?
Baltic	1	?	?	?
South Equatorial Current	3	?	?	?
South Atlantic	50	?	?	?
Mediterranean (Horn et al., 1970)	2.5	540	20	50
Indian Ocean	75	?	?	? (large)
Southwest Pacific (Lee, 1973)	45		<0.01	<0.5
Kuroshio System (Wong et al., unpublished data)	10	14	3.8	38
Rest of northwest Pacific	30		?	? (lower)
Northeast Pacific (Wong et al., unpublished data)	40	3	0.4	16
Arctic (Wong et al., unpublished data)	13		?	? (low)
Antarctic	10		?	?
Area accounted for	120 (33%)			197
TOTAL AREA	361			318[a]

[a] Assuming density 0.5 mg/m^2 for areas unaccounted.

54   PETROLEUM IN THE MARINE ENVIRONMENT

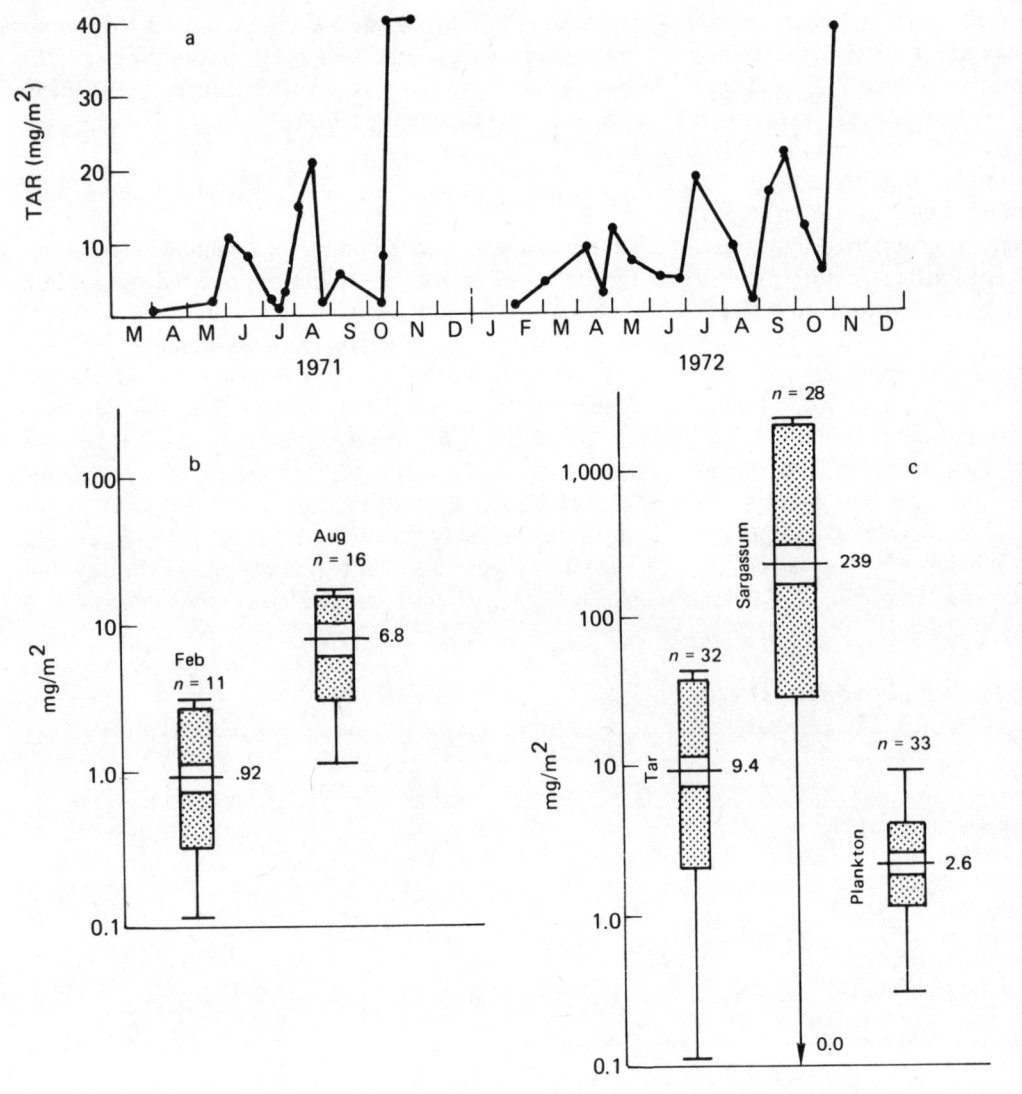

FIGURE 3-6 (a) Time series of tar quantities collected at Station S, 20 mi southeast of Bermuda. (b) Statistics associated with two series of replicate tows. Range, standard deviation of a sample, and standard deviation of the mean are shown. (c) Quantities of tar compared with quantities of *Sargassum* and daylight surface zooplankton.

to estimate the reliability of the sampling. Within each series the amount in a sample appeared to be lognormally distributed with standard deviation approximately 0.5 times the mean value. In the absence of any other statistical data, this magnitude of confidence limits may be placed on other data for pelagic tar concentrations obtained by neuston nets. Thus, the observed seasonal fluctuations between less than 0.1 mg/m² (April 1971) and 40 mg/m² (November 1971) are significant and may be representative of the general "patchiness" of tar found in the open ocean. Comparable variation has been observed on cruise tracks in other parts of the world oceans. (Polikarpov et al., 1971; C. S. Wong, D. Green, and W. J. Cretney, personal communication).

There is a suggestively strong correlation between the observation of high levels of pelagic tar and known tanker routes of high traffic. This is shown in Table 3-3.

The Mediterranean would be expected to have an anomalously low pelagic tar density because of its relatively small size: A tar lump has a greater chance of becoming stranded on a beach than of disintegrating naturally on the sea surface. Girotti (1968) estimated that 30 percent of the oil spilled in the Mediterranean is stranded on beaches.

Distinguishing tar lumps formed from natural crude oil seeps and those resulting from tanker operations is clearly important, but criteria for such distinctions are still being developed. We have already mentioned the presence of C_{30}–C_{40} paraffinic waxes as being indicative of crude oil sludge (hence tanker) origin, and the presence of relatively large amounts of iron as being strong evidence for anthropogenic sources.

If the input of petroleum to the world oceans from natural seeps is about 10 percent of the total, one might expect to find some tar lumps of natural origin in areas

TABLE 3-3 Relation between Occurrence of Spilled Oil and Occurrence of Tar Concentration

Location	Area (10^{12} m^2)	Spilled Petroleum (mg/m^2/yr)	Tar Supply Rate (mg/m^2/yr)[a]	Tar Found Mean (mg/m^2)
North Atlantic	33	17.45	6.13	5.0
Mediterranean	2.5	108	38	20
Kuroshio System	10	33	11.6	3.8
Northeast Pacific	40	0.74	0.26	0.4
Southwest Pacific	45	<0.05	<0.02	<0.005

[a] Based on an average of 35 percent of spill oil becoming tar lumps.

where there are substantial seeps. At present, the surveys of tar quantities are too scattered to permit any specific correlation, but multiparameter analysis of oil from various known seeps and from tar lumps collected in nearby regions might permit some significant correlations to be made. There is reason to believe, however, that since natural seeps contribute crude oil at a slow, more or less steady rate (Allen *et al.,* 1970), the resulting oil slick would be expected to be quite thin and the tar lumps (if any) formed from it would be expected to be of different character from those formed by tanker residues. For example, Ludwig and Carter (1961) report gas bubbles in tar formed from natural seeps. As mentioned above, aggregation of small lumps does not seem to be a significant mode of tar-lump formation, and thus the large pelagic tar lumps collected by most workers would seem to be most likely the result of tanker operations and possible bilge pumping.

Modification of tanker practice to emulsify wastes before dumping would tend to decrease the amount of pelagic tar but increase the amounts in the subsurface water column. Whether the degradation rate to CO_2 would be enhanced by such a change, or whether the sublethal effects on the ocean ecosystem would be increased, are both open to question at present.

WATER COLUMN

In addition to tar lumps, petroleum residues also enter the dissolved and particulate hydrocarbon pool in the deeper oceans. Published data on this is summarized in Table 3-4. Conflicting results and controversy about methods makes any quantitative estimate very uncertain at this time. The surface values in the open ocean appear to be less than 10 μg/liter, and subsurface values are much lower.

Biogenic hydrocarbon input may be comparable: approximately 3 million tons per year (Revelle *et al.,* 1971) to 10 million tons per year (Button, 1971). At the low concentrations found, distinction between biogenic and petroleum hydrocarbons is still beyond the state of the art. However, indirect inference from correlations with phytoplankton may give some information (Zsolnay, 1973b). Of course, a great deal of further field and analytical work would be required to verify this.

SEDIMENTS

Total hydrocarbon concentration (including biogenic compounds) in surface sediment samples determined by a variety of techniques, cover the 0.1–12 ppt range (usually < 1 ppt) in highly polluted coastal areas, less than 100 ppm (usually < 70 ppm) in unpolluted coastal areas and deep marginal seas or basins, and 1–4 ppm (including about 90 percent biogenic) in deep sea areas. Detailed sediment data and references are given in Table 3-5.

CONCLUSIONS

1. Slicks and tar lumps are transient conditions and do not represent the ultimate fate of petroleum spilled on the sea.
2. The ultimate fate is one of the following: (a) evaporation and decomposition in the atmosphere, (b) dispersal in the water column, (c) incorporation into sediments, (d) oxidation by chemical or biological means to CO_2.
3. The standing crop of petroleum-like material in the form of slicks and floating lumps is of the order of a year's input. Tar stranded on rocky shores may have a much longer lifetime.
4. Crude oil and some petroleum products transported by sea can easily form tar. Occurrence of pelagic tar correlates with intensity of tanker traffic in different regions of the ocean.

RECOMMENDATIONS

1. Systematic monitoring of the sea surface, the at-

TABLE 3-4 Hydrocarbons in the Water Column

Location	Depth (m) Collection	Bottom	Concentration (µg/liter)	No. of Samples	Comments	Reference
Bedford Basin, Nova Scotia	<60	200?	1-60	160		Keizer and Gordon, 1973
Narragansett Bay, R.I.	0.2 (microlayer)	10	8.5 5.9	1 1		Duce et al., 1972
Woods Hole Harbor	1	?	11	1		Stegeman and Teal, 1973
Gulf of St. Lawrence	1	20-200	1-15	90		Levy, personal communication
	75	20-200	1-5	90		
	150	20-200	1-10	90		
New York Bight	1	30	1-21	6		Brown et al., 1973
Gulf of Venezuela	1	60	50	1		Brown et al., 1973
Outside Galveston Bay, Tex.	1	15	8	1		Brown et al., 1973
East Bay, La.	3	5	0.2	1	n-alkanes only	Parker et al., 1972
15 mi off Corpus Christi, Tex.	3	20	0.1	1		Parker et al., 1972
15 mi SW Point Au Fer, La. (28°39'N, 19°30'W)	1	20	0.63	1	near burning oil rig	Parker et al., 1972
Mediterranean: Corsica, Italy, area	1	100	170	1		Monaghan et al., 1973
Off Sardinia	1	2,000	195	1		Monaghan et al., 1973
Gulf of Sidra	1	200	4	1		Monaghan et al., 1973
Baltic Sea	surface	?	300-1,000	?	oil products	Simonov and Justchak, 1970
Baltic Sea	1-200		relative values	55	saturated and nonaromatic	Zsolnay, 1973a
Baltic Sea (Gotland Seep)	20	230	48	1		Zsolnay, 1972
	70	230	58	1		
	110	230	58	1		
	150	230	58	1		
	200	230	64	1		
Baltic Sea	1-200		0-50	40	mostly saturated aromatics	Zsolnay, personal communication
Open ocean near Bermuda (similar for Western Atlantic and Caribbean)	surface		~6			Brown et al., 1973; Keizer and Gordon, 1973
	10		~3			
	1,000		~1			
	>2,000		≪1			
Off coast of France	0-50		46-137	3	gas chromatographic and mass spectrographic data	Barbier et al., 1973
Off coast of West Africa	50-4,500		10-95	5		Barbier et al., 1973
Off coast of West Africa	4-35	15-1,000	relative values	58	nonaromatic hydrocarbons correlated with chlorophyll	Zsolnay, 1973b

mosphere, the subsurface water column and, sediments should be more synoptic. Both surveys over large portions of the world oceans and time-series data at a single station are badly needed to form a reliable model of the fate of petroleum in the marine environment.

2. In particular, data on the rate of sedimentation for petroleum residues in both open ocean and coastal areas is badly needed since no reliable measurements have yet been made. The form, concentration, and microdistribution of petroleum residues and biogenic hydrocarbons in surficial sediments should be investigated; the interaction of clastic sediment particles and petroleum should be studied with the aim of understanding the uptake of petroleum residues by naturally occurring sedimentary materials: Transport of petroleum residues along the sediment-water interface is poorly understood and of great importance for predicting the impact of spills in coastal areas, particularly in subtidal environments. Incorporation of petroleum into marsh and tidal flat sediments is in particular need of study since this seems to be a zone of considerable ecological impact.

3. Amounts and types of hydrocarbons in the atmosphere in relatively unpolluted areas, such as the windward side of islands and the open ocean, should be obtained.

TABLE 3-5 Hydrocarbons in Sediments

Location	Sample Depth (cm)	Water Depth (m)	Concentration (ppm, dry wt.)	Number of Samples	References
Buzzard's Bay, Mass.					
Wild Harbor River (over 2½-year period)	top 10 sand and clay	<3	250–1,650	~180	Blumer and Sass, 1972b
Station 37, subtidal unpolluted	top 10	11	38–70	10–20	
Silver Beach (over 2½-year period)	top 10, sand and clay	<2	500–12,000	~60	
Mississippi Coastal Bog	—	0–1.5	350	1	Sever et al., 1972
Narragansett Bay, R.I.					
Mouth of Providence River (Head of Bay)	top 8–10	?	820–3,560	2	Farrington and Quinn, 1973
Middle of Bay		?	350–440	2	
Mouth of Bay		?	50–60	2	
Vineyard Sound, Mass.	top 7	6–9	1.7 n-alkanes only	1	Clark and Blumer, 1967
Chedabucto Bay, Nova Scotia (over 2-year period)	top 5	3	34–420	7	Scarratt and Zitko, 1972
		12	11–1,240	7	
Coast of France					
Vicinity Le Havre	surface	subtidal	450	1	Tissier and Oudin, 1973
Seine Estuary			33	1	
Vicinity Le Havre			920	1	
Bay of Veys			38	1	
Port Valdez, Alaska	surface	subtidal	0.5–2.5 (C_{16}–C_{28} only)	?	Kinney (in press)
Gulf of Batabano, Cuba	?	?	15–85	10	Meinschein, 1969
Orinoco Delta, Venezuela	?	?	27–110	10	Meinschein, 1969
Gulf of Mexico, open ocean[a]	?	?	12–63	10	Meinschein, 1969
Mediterranean, open ocean[a]	?	?	29	1	Meinschein, 1969
Carioca Trench	?	deep	56–352	16	Meinschein, 1969
SE Bermuda to the base of rise near Hudson Canyon	top 5 of sediment	>3,000	1–4	5	Farrington and Madeiros (in press)

[a] Deep marginal seas of basins.

4. Analyses of concentrations in the water column down to 1,000 m are also important for estimating more accurately the pool of petroleum residues in the sea. In this connection, distinction between anthropogenic and natural (biogenic or seep) sources is of political importance for regulation of tanker and other ship disposal practices.

5. Although tar lumps appear to be a small fraction of the total hydrocarbons on the ocean, they are easily collected and analyzed; again, distinction between anthropogenic and natural sources is an important consideration.

6. Experimental studies of petroleum weathering in a natural marine environment are needed, particularly those that go beyond the initial dispersal of slicks from a spill. In this connection, a more systematic study of the photochemical processes acting on spilled oil, as well as the evaporation, dissolution, and emulsification processes, is required.

Biological

MICROBIAL DEGRADATION

In this section heavy reliance has been placed on two comprehensive reviews compiled within the last year (Friede *et al.,* 1972; Ahearn and Meyers, 1973). Most earlier papers may be found cited there, in the bibliography of Moulder and Varley (1971), or in the reviews of Karczewska (1972) and ZoBell (1969, 1971, 1973).

The rates of microbial degradation of fossil hydrocarbons and derivatives in the marine environment vary with the chemical complexity of the crude, the microbial populations, and many of the environmental conditions (Friede *et al.,* 1972; ZoBell, 1973). It must be emphasized that with this multivariable system it is impossible to predict with either ease or accuracy the rate of microbial oil removal. Few reliable field measurements have been made in the marine environment (Blumer *et al.,* 1973; Robertson *et al.,* 1973); laboratory experiments, in which conditions are optimal for oxidation (e.g., Figures 3-7 and 3-8) can only give some indication of maximum rates (Friede *et al.,* 1972; Liu, 1973; Bartha and Atlas, 1973; Kator, 1973; ZoBell, 1973). Even under laboratory conditions, the various fractions of oil or oil products will disappear at rates that can be measured on a time scale of weeks in some instances (Floodgate, 1972, 1973) and that are immeasurably slow in others (Johnson, 1970). Environmental stresses such as temperature and salinity changes, wave action, and sunlight not only directly affect the growth and metabolism of the microorganisms but also alter the physical state (e.g., emulsification) and ultimately the chemical nature (e.g., oxidation) of the hydrocarbons.

The hydrographical and chemical condition into which the oil is deposited varies from place to place. For example, the temperature varies over a range of more than 30 °C; inorganic nutrients range from the high values found in estuaries to very low values in parts of the open ocean. Typical values for seawater are 0.1–1.0 mg/liter N, 0.01–0.1 mg/liter P.

Temperature increases may accelerate growth rates, thereby increasing biodegradation (Friede *et al.,* 1972; ZoBell, 1973). A rise in temperature also increases the rate of evaporation of more volatile components, some of which are degradable and some of which are toxic (Atlas and Bartha, 1972b; see also previous section). Viscosity is lower at higher temperatures, thereby increasing the chance of emulsification and increasing the surface area available for microbial activity and solubility (ZoBell, 1973). Temperature de-

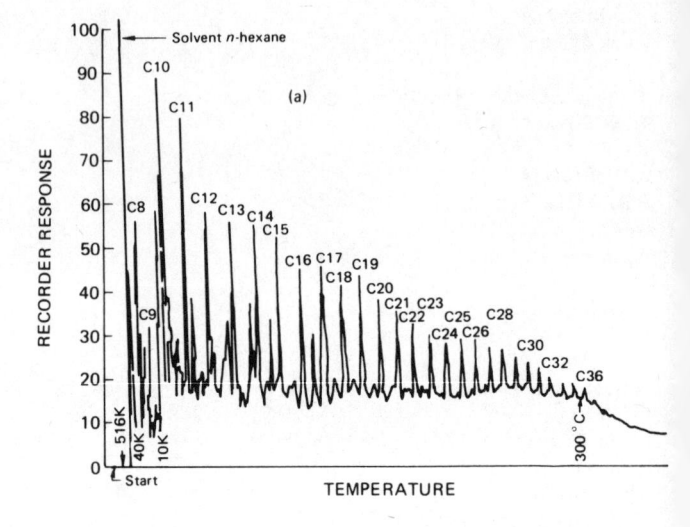

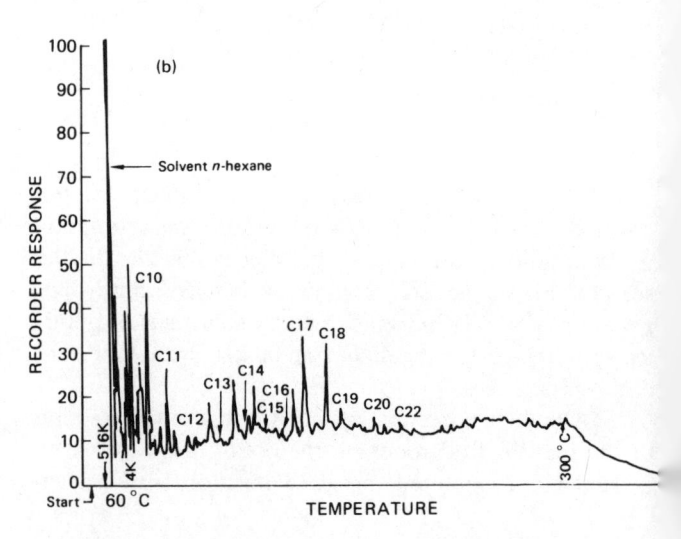

FIGURE 3-7 Gas chromatogram of Canadian crude oil (a). Chromatogram of same oil after the growth of culture CMOL for 2 days (Liu, 1973).

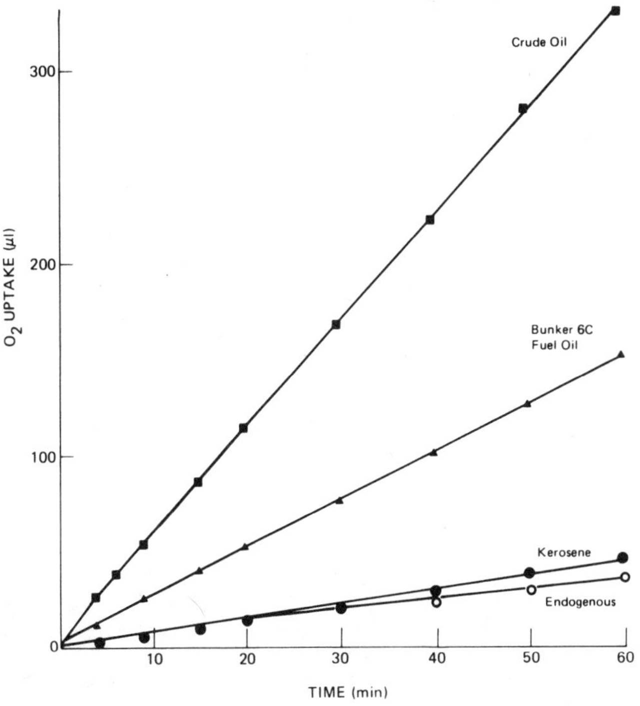

FIGURE 3-8 Oxidation of various petroleum products by culture CMOL. Each Warburg flask contained 1 ml of 0.05 M phosphate buffer (pH 7.0), 1 ml of cell suspension (8.8 mg dry weight), and 100 μm of the substrate. The total fluid volume of each flask was 3.2 ml (Liu, 1973).

creases may not necessarily reduce the overall rate of microbial biodegradation significantly if special psychrophilic cultures develop (Robertson et al., 1973; Traxler, 1973) (Figure 3-9).

Oxygen content is probably always sufficient for degradation of oil at the surface layer and in the upper water column in the open ocean (Friede et al., 1972). The degree of turbulence directly affects the availability of oxygen, as well as the physical dispersion and emulsification of the oil. If the water or sediments become anoxic, then rates of biodegradation will be markedly reduced (Davis, 1967).

Nitrogen and phosphorus concentrations strongly influence the rate of oxidation in laboratory experiments (Figure 3-10) (Gunkel, 1967, 1968; Atlas and Bartha, 1972a). These nutrients may more commonly be limiting in the open oceans than in inshore regions.

Numerous other factors influence biodegradation, for example, presence of sufficient hydrocarbon substrate to develop a viable culture, presence of alternative carbon sources and microbial predators (Gunkel, 1968; Friede et al., 1972), but data are generally insufficient to precisely determine in situ effects on microbial oil utilization.

It might be hoped that the many investigations already carried out would provide an estimate of the rate at which oil might be degraded at sea. Unfortunately, the field and laboratory conditions that have been used, the concentrations and types of oil employed, the multiplicity of ways of measuring activity, and the differing methods of evaluation of the data make generalization difficult and probably misleading. Published estimates of rates based on laboratory studies are probably much higher than would be expected to be found in most of the marine environment. In particular, the often-quoted estimate of 36-350 g/m^3/year (ZoBell, 1964), while it may apply to heavily polluted harbors, is many orders of magnitude too high for the open sea. For example, Robertson et al. (1973) measured oxidation rates of n-dodecane, one of the most readily oxidized hydrocarbons, in situ to be a factor of 1,000 lower (Figure 3-11).

Degradation of crude oil on a rocky shoreline takes many years (Blumer et al., 1973). Additional laboratory and field experiments using carefully evaluated procedures are necessary before any serious advance can be made on this problem. In particular, it is expected that degradation rates for hydrocarbons emulsified or associated with suspended particulates may be more rapid than for hydrocarbons in true solution.

The only substances known with certainty to be produced in the marine environment as a result of microbial digestion, and whose concentration has been measured in the field, are microbial tissue (Gunkel, 1968) and carbon dioxide (Robertson et al., 1973). However, numerous intermediates and end products are known to accumulate in laboratory experiments (Friede et al., 1972; Klug and Markovetz, 1971). Some of these substances, for example aliphatic alcohols, acids, and equivalent aromatic derivatives, may be disruptive to chemotaxis by marine life (Mitchell et al., 1972; Zafiriou, 1972). The microorganisms that digest oil may be pathogenic to or competitively exclude other marine organisms. Some oil-degrading organism could produce toxins (Traxler, 1973). These effects would be most obvious when the numbers of organisms increase suddenly and considerably as a result of an oil spill, but they are difficult to separate from the toxicity of the oil itself, or from stress that lowers resistance to infection. However, little direct evidence either for or against these phenomena has been published. Further discussion of effects is presented in Chapter 4.

A small amount of information indicates that certain bacterial strains may store hydrocarbons in vacuoles (Finnerty et al., 1973). Since some marine animals feed on bacteria, this is a possible route into the food web.

The process of "seeding and/or fertilization" of oil spills to facilitate biodegradation has not been fully

60 PETROLEUM IN THE MARINE ENVIRONMENT

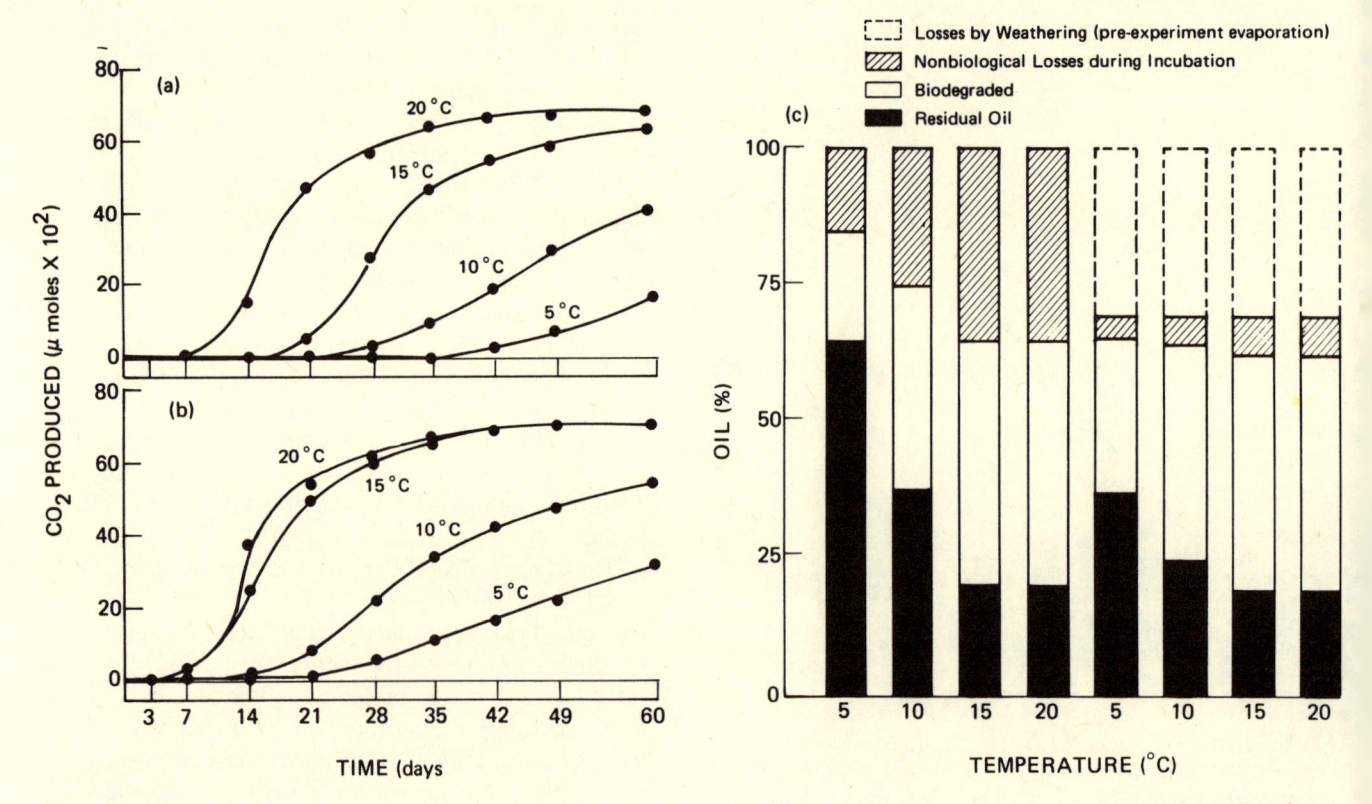

FIGURE 3-9 (a) Mineralization at various temperatures of 1 percent (v/v) fresh Sweden crude oil in seawater collected in winter. (b) As above, but for 0.7 percent (v/v) weathered Sweden crude oil. (c) Losses at various temperatures from 1 percent fresh (left) or from 0.7 percent weathered (right) Sweden crude oil incubated with winter-collected seawater for 60 days (Atlas and Bartha, 1972b).

explored (Miget, 1973), and reports of actual field trials with defined seed systems are available in the literature. There is a potential use for microbial seed cultures, particularly those possessing emulsifying properties, for the treatment of local enclosed conditions such as tanker ballast waters prior to their discharge (Rosenberg and Gutnick, 1973) and for the treatment of bilge waters processed in harbor installations. It seems probably, however, that multiseed stocks of varied metabolic capabilities will be required for different types of oil. At the moment, however, it is not technically feasible to use this method on open waters or beaches.

Uptake by Organisms

Most hydrophobic compounds, such as petroleum and chlorinated hydrocarbons, have very low solubilities in water, generally in the ppm to ppb range (Bohon and Clausen, 1951; McAuliffe, 1966, 1969b; Peake and Hodgson, 1966, 1967). Low-molecular-weight paraffinic and aromatic hydrocarbons have relatively high solubility in water, but because of their volatility are probably lost to the atmosphere from surface slicks. Thus, petroleum can be presented to pelagic organisms as dissolved, dispersed, or as floating tar lumps, whereas benthic organisms are faced with petroleum in the sediment, as well as the dissolved and dispersed phases.

We suggest that hydrocarbons can enter the marine food web by several means. One is through adsorption to particles, both living and dead, followed by ingestion of these particles. A second entrance is through active uptake of dissolved or dispersed petroleum, mainly via the gills. A third means would be passage into the gut of animals that gulp or drink water.

There are in the literature a number of analyses of the hydrocarbon levels that occur in marine organisms without necessarily indicating the mode of uptake. Specific analyses for petroleum hydrocarbons are presented in Table 3-6. We have attempted to exclude biogenic hydrocarbons by subtracting levels reported for control (unpolluted) samples (e.g., Figure 3-12) in those cases where the authors have not already done so. Because of the complexity of petroleum hydrocarbons, the table lists the fraction of the petroleum for which the analysis has been done. This, of course, makes it difficult to compare different analyses quantitatively.

Although analyses of carcinogenic hydrocarbons in marine organisms have been compiled (ZoBell, 1971), it is impossible to distinguish the extent to which these compounds are petroleum derived. Plankton analyses may also be in question due to possible contamination from the sampling nets (Harvey and Teal, 1970).

In a laboratory bioassay simulating an oil slick on tidal waters, mussels incorporated paraffin hydrocarbons from a No. 2 fuel oil and, after removal to fresh seawater, retained them for 5 weeks with a small decrease in hydrocarbon content with time. In a similar situation, No. 5 fuel oil showed considerably less uptake and a rapid loss (Clark and Finley, 1973b). Lee and co-workers (1972a) found that dissolved hydrocarbons (saturates and aromatics), although taken up rapidly by mussels, were also discharged without metabolic breakdown after the mussels were placed in fresh seawater.

Mussels exposed to 10 percent two-cycle-out-board motor effluent (dissolved and emulsified) in seawater showed abnormal respiration (gaping after 24 h) and histopathology (gill tissue damage). Under these conditions, petroleum hydrocarbon uptake, as indicated by paraffin hydrocarbon patterns in the exposed mussels, matched the dissolved/emulsified effluent extract after allowances for the biogenic paraffin content of the controls had been made (Clark et al., 1973b).

The variation of uptake and loss of petroleum hydrocarbons under laboratory conditions (Clark and Finley, 1973b; Lee et al., 1972a) and those found after real-life spills (Clark and Finley, 1973a) are related to the magnitude of the exposure—the amount of the pollutant and the duration, as well as the physical and chemical properties of the pollutant (Stegeman and Teal, 1973).

Once within the organisms there is further distribution within the body, including distribution into compartments from which loss is slow.

One must appreciate that only a small fraction of

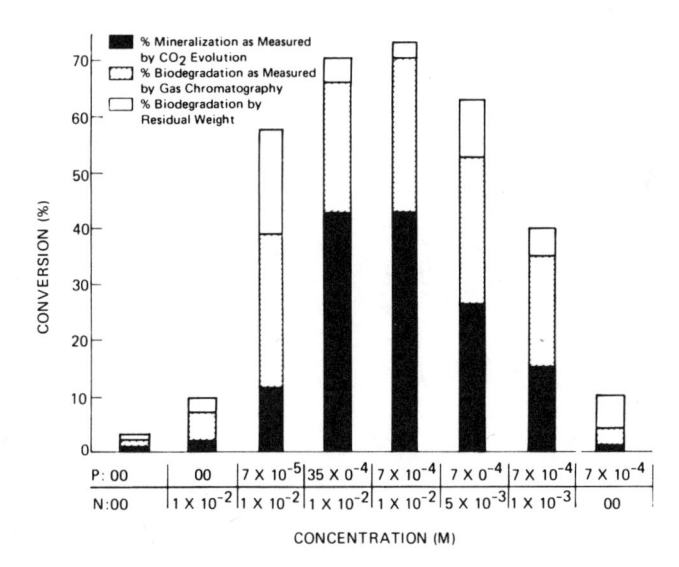

FIGURE 3-10 Conversion of petroleum in seawater supplemented by various concentrations of Na_2HPO_4 and KNO_3 during 18 days of incubation (Atlas and Bartha, 1972a).

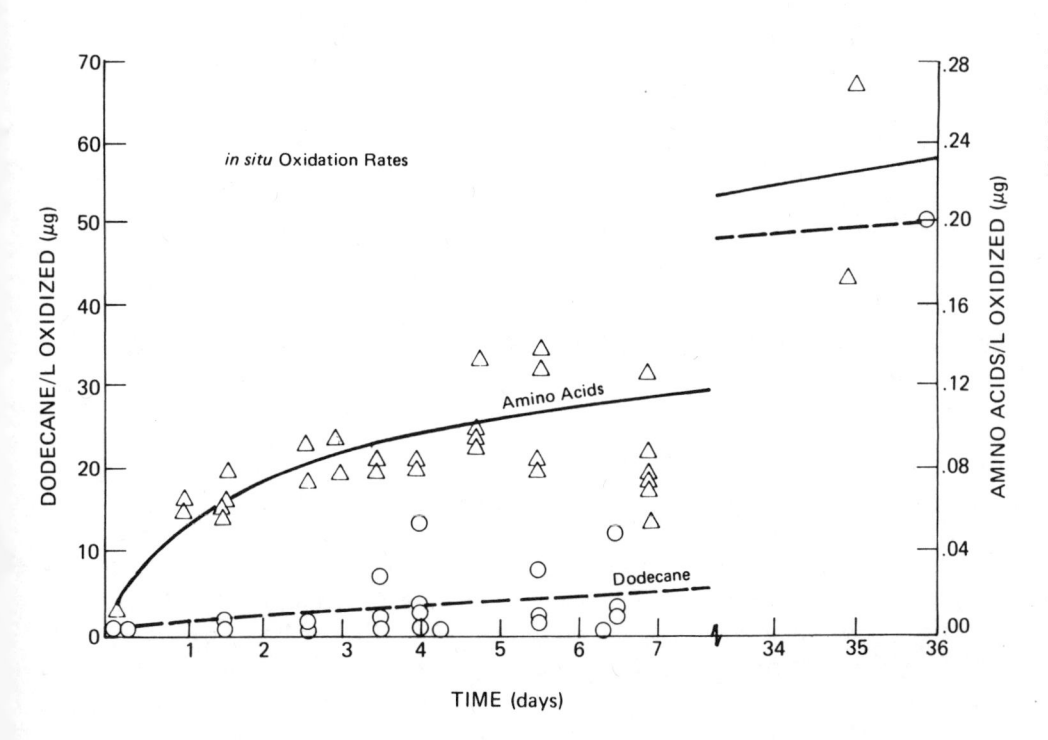

FIGURE 3-11 Carbon dioxide liberated from radioactive dodecane and from an amino acid mixture incubated in Port Valdez (Robertson et al., 1973).

TABLE 3-6 Petroleum Hydrocarbon Levels in Marine Macroorganisms

Organisms	Area Type[a]	Hydrocarbon Type	Estimated Hydrocarbon Amount ($\mu g/g$)	Reference[b]
Macroalgae				
Fucus	4	Bunker C[c]	40 dry	Clark et al., 1973b
Enteromorpha	4	No. 2 fuel oil	429 wet	Burns and Teal, 1971
Sargassum	1	$C_{14}-C_{30}$ range	1–5 wet	Burns and Teal, 1973
Higher plants				
Spartina	4	No. 2 fuel oil	15 wet	Burns and Teal, 1971
Molluscs				
Modiolus, mussel	4	No. 2 fuel oil	218 wet	Burns and Teal, 1971
Mytilus, mussel	4	No. 2 fuel oil[c]	36 dry	Clark and Finley, 1973a
Mytilus	4	Bunker C[c]	10 dry	Clark and Finley, 1973a
Mytilus	4	Bunker C, aromatics	74–100 wet	Zitko, 1971
Mytilus	3	n-$C_{14}-C_{37}$[c]	9 dry	Clark and Finley, 1973a
Mya, clam	4	No. 2 fuel oil	26 wet	Blumer et al., 1970b
Pecten, scallop	4	No. 2 fuel oil	7 wet	Blumer et al., 1970a
Littorina, snail	4	Bunker C, aromatics	46–220 wet	Zitko, 1971
Mercenaria, clam	3	$C_{16}-C_{32}$ range	160 dry	Farrington and Quinn, 1973
Crassostrea, oyster	2	polycyclic aromatics	1 wet	Cahnmann and Kuratsune, 1957
Crustacea				
Hemigrapsus, crab	4	Bunker C[c]	8 dry	Clark et al., 1973a
Mitella, barnacle	4	Bunker C[c]	8 dry	Clark et al., 1973a
Lady crab	3	$C_{14}-C_{30}$	4 wet	Bowen, 1971
Plankton	2	benzopyrene	0.4 wet	Mallet and Sardou, 1964
Sargassum shrimp	1	$C_{14}-C_{30}$	3 wet	Burns and Teal, 1973
Lepas, barnacle	1	$C_{14}-C_{30}$	6 wet	Bowen, 1971
Portunus, crab	1	$C_{14}-C_{30}$	34 wet	Burns and Teal, 1973
Planes, crab	1	$C_{14}-C_{30}$	11 wet	Burns and Teal, 1973
Fish				
Fundulus, minnow	4	No. 2 fuel oil	75 wet	Burns and Teal, 1971
Anguilla liver, eel	4	No. 2 fuel oil	85 wet	Burns and Teal, 1971
Smelt	3	benzopyrene	0–5 dry	Lee et al., 1972b
Flatfish	2	$C_{14}-C_{20}$	4 wet	Bowen, 1971
Flying fish	1	$C_{14}-C_{20}$	0.3 wet	Bowen, 1971
Sargassum fish	1	$C_{14}-C_{20}$	1.6 wet	Burns and Teal, 1973
Pipefish	1	$C_{14}-C_{20}$	8.8 wet	Burns and Teal, 1973
Triggerfish	1	$C_{14}-C_{20}$	1.7 wet	Burns and Teal, 1973
Birds				
Herring gull, muscle	4	No. 2 fuel oil	535 wet	Burns and Teal, 1971
Echinoderm				
Asterias, starfish	4	Bunker C, aromatics	20–147 wet	Zitko, 1971
Luidia, starfish	2	$C_{14}-C_{20}$	3.5 wet	Bowen, 1971

[a] Area types: (1) oceanic; (2) chronic pollution, coastal; (3) chronic pollution, harbor; (4) single spill.
[b] See reference for method of distinguishing petroleum hydrocarbons from native biogenic hydrocarbons and discussions concerning techniques.
[c] n-Alkanes only.

petroleum has a pronounced odor or taste. Thus, while observations of "tainting" (Simpson, 1968) may indicate high-level contamination by certain petroleum pollutants, they are inconclusive in proving that oil has not been incorporated (Blumer, 1971).

Some of the ingested material is not absorbed but instead passes directly through the animals and is lost in the feces. Following the *Arrow* incident in Chedabucto Bay, plankton were observed to ingest large quantities of small drops of Bunker C oil and eliminate them in the form of fecal matter (up to 7 percent Bunker C oil by weight) (Conover, 1971). No chemical analysis of the fecal matter or of the whole copepods was reported that might have provided some indication of whatever degradation or partitioning, if any, of the oil took place. The plankton always voided the small "oil" particles within 24 h and showed no signs of stress when viewed under a dissecting microscope. Conover (1971) went further to suggest that under the conditions observed in Chedabucto Bay, the plankton could graze as much as 20 percent of the oil particles (less than 1 mm in diameter) in the water column and sediment them in their denser than seawater feces.

Parker (1970) also demonstrated the presence of considerable quantities of oil in the guts of copepods and barnacle larvae and in their fecal pellets. The fact that the oil passes unchanged into the fecal material is of considerable interest since oil from a slick can be grazed by the plankton and the ingested oil precipitated in the feces. Parker (1971) calculated that copepods (*Calanus finmarchicus*) could encapsule up to 1.5×10^{-4} g of oil per day per individual. For example, a population of 2,000 individuals/m³ covering an area of 1 km² to a depth of 10 m could remove as much as 3 tons of oil daily if its concentration was 1.5 ppm or greater. Fecal pellets can then be eaten by other members of the food web.

Alyakrinskaya (1966) found that *Mytilus galloprovincialis* in the Black Sea could tolerate high concentrations of oil (undefined type of oil; up to 20 ml/liter); during filtration of oil-polluted water, the molluscs formed pseudofeces from oil connected by mucous—to a degree comparable with transferring the oil to larger, denser particles as Conover and Parker have suggested for copepods.

Equilibration of hydrocarbons can occur between organisms and the seawater that passes over their gills or other membranes exposed to seawater. This may be the most important route for most aquatic animals since they process such large amounts of water during food collection and respiration. One can calculate from the hydrocarbons measured in coastal waters (Stegeman and Teal, 1973; Brown et al., 1973) of 10 μg/liter and a level in food of 10 μg/g (Table 3-6) that an animal would be exposed to more than an order of magnitude larger amount of hydrocarbons in the water processed to obtain oxygen for metabolism of the food than that amount present in the food itself. Stegeman and Teal believe that uptake from the water is the major route by which their oysters accumulated hydrocarbons from the water. In other situations, uptake from sediments could also be important.

Dissolved hydrocarbons are taken up by the gill tissue of the mussel *Mytilus edulis,* followed by transfer of the hydrocarbons to other tissues (Lee et al., 1972a). Studies on uptake of iron by electron microscopy suggest the gill tissue of this mussel has a micellar layer on the surfaces of the gill that is responsible for the absorption of hydrophobic compounds (Pasteels, 1968). Work on the uptake of dissolved hydrocarbons by marine fish also demonstrated the entrance of hydrocarbon through the gills (Lee et al., 1972b). Phytoplankton appeared to absorb hydrocarbon, but there was no transfer of hydrocarbon inside the cell (R. F. Lee, personal communication).

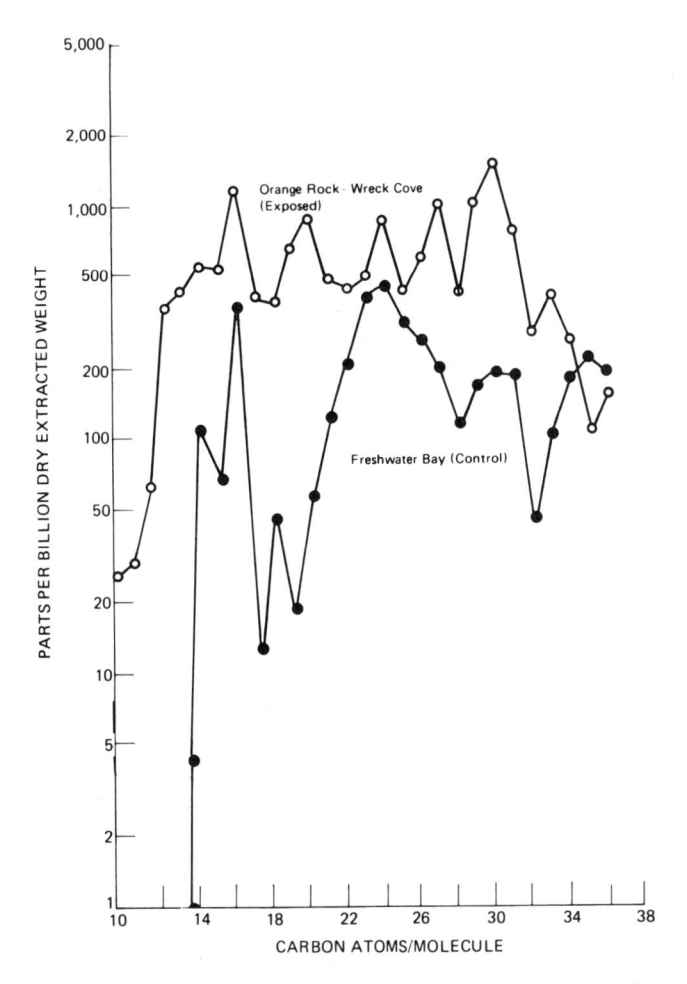

FIGURE 3-12 Paraffin hydrocarbon patterns for mussels (*Mytilus californianus*) exposed to moderate level oil pollution. Exposed mussels collected at Orange Rock, 2 months after the grounding of the *General M. C. Meigs* are compared with control mussels collected at Freshwater Bay (from Clark and Finley, 1973a).

Yevich and Barry (1970) reported on tissue damage brought about by exposure to crude oils and other pollutants; such damage includes sloughing of the epithelium and atypical basal cell hyperplasia of the ciliated inner gills of quahogs (clam, *Mercenaria mercenaria*). The question also arises, then, of what effect the loss of the protective membrane coatings of the gills has on rate of absorption of hydrocarbons from water.

Once hydrocarbons are within the animal, changes in distribution between various tissues is a further indication of equilibration across membranes. Some pathways can be inferred from the distribution of biogenic hydrocarbons such as pristane (Blumer et al., 1964). Mullet caught in Australia have been shown sometimes to acquire a kerosene-like taint that, upon analysis of the tissue using gas chromatography and spectral analysis, appears to be similar, qualitatively and quantitatively, to that of a commercial sample of kerosene (Shipton et al., 1970). The dark meat and the fatty layer adjacent to the skin were more severely tainted than the white meat, and the tainted flesh (fillets) had a higher fat content than those from untainted fish caught at the same time. The fatty acid analyses also showed that the tainted fish had a lower C_{16} and a higher C_{18} acid content than the untainted. Examination of livers by an optical microscope and an electron microscope showed excessive amounts of free fat, typical of fatty infiltration, in the livers of the tainted fish, compared with the untainted mullet (Vale et al., 1970). These observed changes in liver fat, fatty acid composition, and total lipid content can be assumed to be related to the fate of the pollutant, but no direct degradation studies were reported. Fatty liver in higher animals can be caused by petroleum distillates (grouped under the British term of "benzine"; Browning, 1953).

Based on *in vivo* experiments that show hydrocarbons preferentially invade nerve tissue, the hydrocarbon-membrane interaction has been investigated by Roubal (1973) using nonlabeled hydrocarbons and electron paramagnetic resonance probes. *In vitro* experiments using excised spinal cord tissues of coho salmon indicate that hexane and similar hydrophobic compounds (cyclohexane, pentane, octadecane, and hexadecane) are directed away from the nerve membrane surface to sites situated more deeply in the lipid bilayer of the membrane, while the aromatic hydrocarbons (benzene, toluene, xylene, and ethylbenzene) and benzyl alcohol contribute to membrane surface changes. The low-molecular-weight aromatics alter surface organization of nerve membranes more effectively than nonaromatics.

Membrane disorganization by hydrocarbon pollutants may involve (1) alteration in membrane proteins, (2) changes in ion-binding properties of protein and/or phospholipids, and (3) modifications in membrane permeability.

METABOLISM

The metabolic pathways involving oxidases and other enzymes, important in the degradation of aromatic and paraffinic hydrocarbons by mammalian systems, have been well studied (Boyland and Solomon, 1955; Falk et al., 1962; McCarthy, 1964; Diamond and Clark, 1970; Daly et al., 1972). In the case of aromatic hydrocarbons, hydroxylation is followed by conjugation with sulfate or glucose and finally excretion of the water-soluble product. Straight chain hydrocarbons are hydroxylated at the terminal end and further oxidized to the fatty acid that can be broken down by β-oxidation. Highly branched chain hydrocarbons, such as pristane and phytane, are probably oxidized to an acid (e.g., phytanic acid), which can be further oxidized by a combination of α- and β-oxidation (Mize et al., 1969).

How widespread is the presence of these metabolic pathways in a variety of organisms in the sea? Although we cannot give a complete or satisfactory answer to this question, we can discuss the little information available for a few species of organisms from several marine environments. It should be mentioned here that the methodology for this type of work is well worked out, based on extensive research with mammals.

Degradation of sizeable amounts (between 10 and 500 μg) of aromatic and paraffinic hydrocarbons occur in marine fish and some marine invertebrates (Stegeman and Teal, 1973; Lee et al., 1972a,b). Other benthic marine invertebrates, phytoplankton, and some zooplankton, over a period of a month, were unable to oxidize either paraffinic or aromatic hydrocarbons. Several species of copepods were unable to metabolize hydrocarbons but could degrade paraffinic hydrocarbons (R. F. Lee, personal communication). The liver or the liverlike organ in invertebrates, the hepatopancreas, is assumed to be the site of hydrocarbon degradation. Unaltered hydrocarbons are sent to these organs where hydroxylation and other detoxification reactions occur. In those invertebrates where degradation does not occur, perhaps some of the detoxifying microsomal oxidases are missing from the hepatopancreas. Hydroxylated products are found when fish and some crustacea are given aromatic and paraffinic hydrocarbons. The serum lipoproteins appeared to have a role in carrying the hydroxylated hydrocarbons to various tissues from the liver in both fish and lobsters.

We should keep in mind during this discussion that long chain paraffinic hydrocarbons (carbon chain lengths between C_{12}–C_{30}) are a common constituent of marine organisms, although usually accounting for only a few percent of the lipid. However, most aromatic hydrocarbons present in petroleum are not known to be synthesized by marine organisms, though there are reports of biosynthesis of benzpyrenes by freshwater green algae (Borneff and Fischer, 1962). Perhaps half of the hydrocarbons in the sea are due to hydrocarbons manufactured by living organisms, while half are due to petroleum compounds (Button, 1971). This assumption is valid only if the hydrocarbons synthesized by living organisms are metabolized at the same rate as petroleum hydrocarbons. It certainly seems clear from our discussion that many petroleum hydrocarbons are more resistant to degradation than hydrocarbons synthesized by living organisms.

STORAGE

Hydrocarbons are to some extent stored within the bodies of marine animals and either are not metabolized or metabolized only very slowly. The lipids of copepods of the genus *Calanus* contain pristane that increases in relative concentration as the lipids are metabolized during starvation (Blumer et al., 1964). Blumer (1967) has shown that in basking sharks that feed on *Calanus*, the pristane is absorbed through the digestive tract without structural modification and deposited in the liver. Squalene, which is also stored in the shark liver, is apparently synthesized there, since it occurs only in small amounts in zooplankton.

Storage of hydrocarbons in the liver occurs in marine fish and in the hepatopancreas of several invertebrates (Lee et al., 1972a,b; Lee, personal communication). Since the liver and hepatopancreas are generally high in lipid, this observation would be predicted. The gall bladders in fish are also a temporary storage site, although this organ apparently serves mainly as an avenue for discharge. The complex lipoproteins of plasma membranes and organelle membranes of all tissues are also possible storage sites. After various periods of exposure to hydrocarbons, animals capable of metabolizing hydrocarbons, such as fish and lobsters, were transferred to hydrocarbon-free seawater and after 14 days, less than 1 percent of the original hydrocarbon could be detected in the various tissues. Animals that were not degrading the hydrocarbons, such as copepods and mussels, showed retention of about 10 percent of the hydrocarbons after a 14-day discharge period. However, the absolute retention depends on the initial amount of exposure; for high initial concentrations, substantial amounts may be retained even if a large fraction of the original contamination is eliminated. Blumer et al. (1970a,b) found stores of petroleum still present in oysters from the West Falmouth area many months after a spill. Stegeman and Teal (1973) believe oysters may metabolize paraffins and retain a portion of the other petroleum compounds.

Mixed marine zooplankton show less biogenic pristane and more C_{19} and C_{20} olefins as sampling progresses from the Gulf of Maine with its pristane-rich calanoid copepods toward the mid-Atlantic along 42°N (Blumer and Thomas, 1965b). Various isomeric C_{19} monoolefins have been isolated from Gulf of Maine zooplankton (Blumer, 1967), and they, along with pristane (Blumer et al., 1969) and several phytadienes (Blumer and Thomas, 1965a), originate in phytol from phytoplankton ingested by the zooplankton.

Di- and triolefins are found in marine zooplankton but are not present in ancient sediments or in petroleum; therefore, they can serve as a valuable criterion for the distinction between biogenic hydrocarbons derived from aquatic macroorganisms and those hydrocarbons taken up from petroleum pollution; i.e., if di- and triolefins are present in certain rations with the petroleum-contributed hydrocarbons, then the degree of petroleum contamination could be estimated (Blumer et al., 1969).

DISCHARGE

Using the aromatic hydrocarbons benzene, toluene, 1,2,3,4-tetrahydronaphthalene, naphthalene, benzo(a)-pyrene, and methylcholanthrene, as well as the paraffinic mineral oil, heptadecane, Lee et al. (1972a) studied the uptake and eventual discharge of these compounds by marine fish, two benthic invertebrates (mussel and lobster), and several species of zooplankton. As discussed above, copepods and mussels were unable to degrade aromatic hydrocarbons to water-soluble products; nonetheless approximately 90 percent of the accumulated hydrocarbon was discharged in 14 days. Possibly bile salts or some other natural detergents were able to emulsify these hydrocarbons and allow passage through the gut and into the feces or pseudofeces. Fish made water-soluble products from the hydrocarbons, and the main avenue of discharge appeared to be through the urine via the gall bladder and kidney. In mammals aromatic hydrocarbons are also converted to water-soluble products that go through the bile and into the feces and urine (Falk et al., 1962; Diamond and Clark, 1970). The avenue for the discharge of hydrocarbons by the lobster and related invertebrates has not been determined.

Stegeman and Teal (1973) and Anderson (1973) have demonstrated that oysters discharge 90 percent of *n*-paraffins within a few days (Figure 3-13) though they retain aromatic hydrocarbons longer. In contrast, mussels collected from Scripps pier showed a buildup of petroleum hydrocarbons for several days after a fuel oil spill. But three weeks later, none of the material could be found in the mussels (Lee and Benson, 1973). The levels of exposure were very different from that in the oyster experiments but there may also be differences in the ability of these molluscs to discharge hydrocarbons.

Where specific analyses of hydrocarbons in the tissues of molluscs are available (Anderson, 1973), the uptake and discharge seem related to the chemical nature of the compounds. In both the clam *Rangia cuneata* and the oyster *Crassostrea virginica*, aromatic hydrocarbons are taken up by the tissues to a greater extent than the saturated forms. Of the aromatics, naphthalene and alkylnaphthalenes are accumulated most readily, particularly when bivalves were exposed

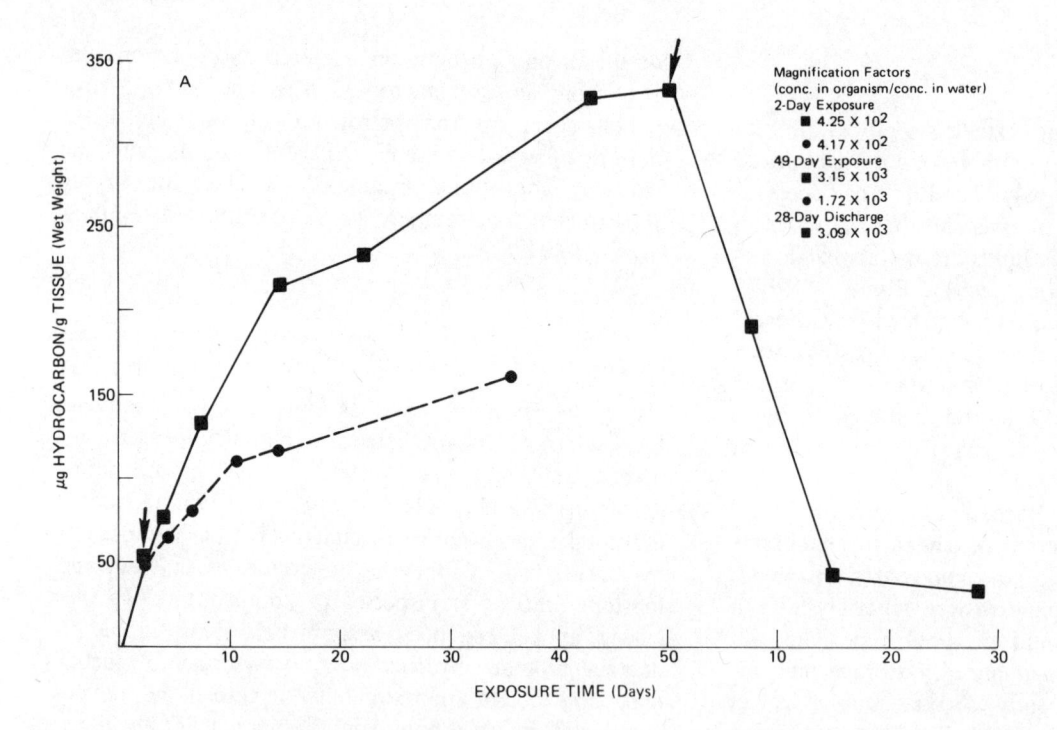

FIGURE 3-13 Uptake and release of petroleum hydrocarbons by high fat content (■) and low fat content (●) oysters, wet weight basis. The concentration of hydrocarbons in the water was 106 µ/liter. At day 50, oysters were transferred to a system with 11 µg/liter hydrocarbons. Each point represents three oysters (Stegeman and Teal, 1973).

to No. 2 fuel oil and Bunker C. While these compounds rapidly enter the tissues, further research is needed to determine the relationship between tissue content and toxicity. Depuration studies (Anderson, 1973) have shown that oysters exposed to crude and fuel oil were free of detectable levels of aromatic and saturated hydrocarbons after maintenance for 24–52 days in clean water.

Food Web Magnification

It is commonly assumed that petroleum in the sea is absorbed by particulate matter and microorganisms. These are then eaten by zooplankton and other animals that accumulate the hydrocarbons, which in higher links in the food web further magnify the hydrocarbon concentrations (Holcomb, 1969). We are using "food web magnification" to mean an increasing concentration of hydrocarbons per weight of tissue or lipid at successively higher trophic levels. In the case of chlorinated hydrocarbons, this magnification is well demonstrated in birds (Risebrough, 1969; Risebrough et al., 1968). But as we have suggested above, in the case of animals with respiratory surfaces in contact with seawater (or fresh water), partitioning between the water and the animal may be the most important avenue for both uptake and loss of hydrocarbons. Burns and Teal (1973) "found no relation between either total amount of pollutant hydrocarbons or ratio of pollutant to natural compounds and the animals' supposed positions in the food chain." The same authors (1971) found an inverse relation between the concentration of hydrocarbons per lipid weight and food web position in organisms from the Wild Harbor marsh. Apparent food chain magnification may more likely be a function of the ability of different species to accumulate hydrocarbons from the water than a function of their position in the food web.

Conclusions

1. Neither a single rate nor a mathematical model for the rate of petroleum biodegradation in the marine environment can be given at present. On the basis of available information, the most that can be stated is that some microorganisms capable of oxidizing chemicals present in petroleum (under the right conditions) have been found in virtually all parts of the marine environment examined.

2. Laboratory experiments have demonstrated that the n-alkane fraction of petroleum is most easily degraded. In oxic marine environments, this type of compound is likely to be degraded in a matter of days or months, depending principally on temperature and nutrient supply. Other fractions are more resistant to microbial action, and the time required for substantial decomposition of the most resistant components of petroleum in the marine environment is probably measured in years to decades.

3. In larger organisms, hydrocarbons are taken up

primarily through the gills or by ingestion of particulate matter. Direct uptake from water through the gills is probably the most important pathway in pelagic environments. For benthic organisms, the sediments may be a more important source.

4. The measured level of petroleum hydrocarbons (after correction for biogenic contributions) in a variety of marine organisms ranges over three orders of magnitude. Available data (1 μg/g to 400 μg/g wet weight) are summarized in Table 3-6.

5. Some organisms (e.g., copepods) can ingest large quantities of petroleum and eliminate it directly as fecal matter without substantial degradation.

6. Some fish and crustaceans metabolize petroleum hydrocarbons within 2 weeks; in plankton and benthic invertebrates, however, metabolism is slow and the pathways are poorly understood.

7. Storage of hydrocarbons, including those from petroleum, occurs in the lipids of many organisms. Biogenic hydrocarbons, particularly di- and triolefins, are often clearly distinguishable from petroleum.

8. Some organisms (e.g., mussels and oysters) can eliminate most petroleum hydrocarbons (but not all) after absorption if placed in unpolluted water.

9. Discharge by vertebrates occurs primarily through the gall bladder and kidney. Paths of discharge for invertebrates are not well established.

10. There is no evidence for food web magnification in the case of petroleum hydrocarbons in the marine environment. On the contrary, evidence is strongest that direct uptake from the water or sediments is more important than from the food chain, except in special cases.

RECOMMENDATIONS

1. Microbial degradation studies should be conducted under environmental conditions comparable to those encountered in the field. New techniques should be developed to permit more accurate systematic observation or control of relevant environmental parameters (e.g., temperature, nutrients, oil composition).

2. New techniques should be developed for work in the field, and extensive field studies executed. At present, because laboratory results cannot be extrapolated in a meaningful way to the marine environment, it is imperative that well-designed and rigorously prosecuted field studies be conducted on the beach, in the open ocean, and in estuaries.

3. Research into uptake of petroleum hydrocarbons by marine organisms is divided into three basic areas:

   a. Absorption of dissolved components through the gill membranes into the circulatory system; determination of the rate and type of chemical fractionation of this pathway, compared with the digestive pathway;

   b. Physical phenomenon of coating, as in the accumulation of oil on the surface of gill tissue; and

   c. Uptake of ingested material through the digestive system (soluble components, material adsorbed on particulates, oil droplets, oil in the lipid of detritus, etc.).

4. Metabolic fate of hydrocarbons present in petroleum should be studied in a wide variety of marine organisms. Mechanisms of detoxification, metabolism, and elimination and rates of uptake and turnover should be determined experimentally.

5. Incorporation of hydrocarbons into fecal material and the associated transfer to deep sea organisms and sediments should be investigated.

6. The effect of "sublethal" environmental variables such as temperature, hydrocarbon concentration, and nature of hydrocarbon dispersion on uptake, metabolism, and elimination, as well as any transfer either up or down the food chain, should be studied at hydrocarbon levels *low* enough to prevent any toxicity or ecosystem disruption. Residence times of petroleum hydrocarbons in the lipid pools of various organisms should be obtained.

References

Adlard, E. R., L. F. Creaser, and P. H. D. Matthews. 1972. Identification of hydrocarbon pollutants on seas and beaches by gas chromatography. Anal. Chem. 44:64–73.

Ahearn, D. G., and S. P. Meyers, eds. 1973. The microbial degradation of oil pollutants. Workshop held at Georgia State University, Atlanta, December 1972. Publ. No. LSS-SG-73-01. Center for Wetland Resources, Louisiana State University, Baton Rouge. 322 pp.

Allen, A. A. 1969. Statement to U.S. Senate Interior Committee, Subcommittee on Minerals, Materials and Fuels. May 20, 1969.

Allen, A. A., R. S. Schlueter, and P. J. Mikolaj. 1970. Natural oil seepage at Coal Oil Point, Santa Barbara, California. Science 170:974–977.

Alyakrinskaya, I. O. 1966. Behavior and filtering ability of the Black Sea *Mytilus galloprovincialis* on oil polluted water. Zool. Zh. 45(7):998–1003; also in Biol. Abstr. 48(14):6494.

Anderson, J. W. 1973. Uptake and depuration of specific hydrocarbons from fuel oil by the bivalves *Rangia cuneata* and *Crassostrea virginica*, pp. 690–708. *In* Background papers, workshop on petroleum in the marine environment. National Academy of Sciences, Washington, D.C.

Asthana, V., and J. I. Marlowe. 1970. Oil contamination and the coast. Unpublished report of Atlantic Oceanographic Laboratory, Dartmouth, Nova Scotia.

Atlas, R. M., and R. Bartha. 1972a. Degradation and mineralization of petroleum in sea water. Biotech. Bioeng. 14:297–318.

Atlas, R. M., and R. Bartha. 1972b. Biodegradation of petroleum in seawater at low temperature. Can. J. Microbiol. 18:1851–1855.

Attaway, D., J. R. Jadamec, and W. McGowan. 1973. Rust in floating petroleum found in the marine environment. U.S. Coast Guard, Groton, Conn. Unpublished manuscript.

Baier, R. E. 1970. Surface quality assessment of natural bodies of water. Proc. Conf. Great Lakes Res. 13:114–127.

Baier, R. E. 1972. Organic films on natural waters: Their retrieval, identification, and modes of elimination. J. Geophys. Res. 77:5062–5075.

Baker, E. G. 1967. A geochemical evaluation of petroleum migration and accumulation, pp. 299–329. *In* Fundamental Aspects of Petroleum Geochemistry. Elsevier, Amsterdam.

Barbier, M., D. Joly, A. Saliot, and D. Tourres. 1973. Hydrocarbons from sea water. Deep Sea Res. 20:305–314.

Bartha, R., and R. M. Atlas. 1973. Biodegradation of oil in seawater: Limiting factors and artificial stimulation, pp. 147–152. *In* Ahearn and Meyers, eds., *op. cit.*

Beerstecher, E. 1954. Petroleum Microbiology. Elsevier, New York and Amsterdam.

Berridge, S. A., R. A. Dean, R. G. Fallows, and A. Fish. 1968a. The properties of persistent oils at sea. J. Inst. Pet. 54:300–309.

Berridge, S. A., M. T. Thew, A. G. Loriston-Clarke. 1968b. The formation and stability of emulsions of water in crude petroleum and similar stocks. J. Inst. Pet. 54:337–357.

Blokker, P. C. 1964. Spreading and evaporation of petroleum on water. Proceedings of the 4th International Harbour Conference, Antwerp, Vol. 5, pp. 911–919. Ingenieursvereinigung, Antwerpen.

Blumer, M. 1967. Hydrocarbons in digestive tract and liver of a basking shark. Science 156:390–391.

Blumer, M. 1970. A polyunsaturated hydrocarbon in the marine food web. Mar. Biol. 6:226–235.

Blumer, M. 1971. Scientific aspects of the oil spill problem. Environ. Affairs 1:54–73.

Blumer, M., and J. Sass. 1972a. The West Falmouth oil spill. Woods Hole Oceanographic Institution Tech. Rep. 72:19. Woods Hole, Mass. 125 pp.

Blumer, M., and J. Sass. 1972b. Oil pollution: persistence and degradation of spilled fuel oil. Science 176:1120–1122.

Blumer, M., and D. W. Thomas. 1965a. Phytadienes in zooplankton. Science 147:1148–1149.

Blumer, M., and D. W. Thomas. "Zamene," isomeric C_{19} mono-olefins from marine zooplankton, fishes and mammals. Science 148:370–371.

Blumer, M., M. M. Mullin, and D. W. Thomas. 1964. Pristane in the marine environment. Helgoländer Wiss. Meersunters. 10:187–201.

Blumer, M., J. C. Robertson, J. E. Gordon, and J. Sass. 1969. Phytol-derived C_{19} di- and triolefinic hydrocarbons in marine zooplankton and fishes. Biochemistry 8:4067–4074.

Blumer, M., G. Souza, and J. Sass. 1970a. Hydrocarbon pollution of edible shellfish by an oil spill. Mar. Biol. 5:195–202.

Blumer, M., J. Sass, G. Souza, H. Sanders, F. Grassle, and G. Hampson. 1970b. The West Falmouth oil spill. Woods Hole Oceanographic Institution Tech. Rep. 70–44. Woods Hole, Mass. 32 pp.

Blumer, M., L. L. Sanders, J. F. Grassle, and G. R. Hampson. 1971. A small oil spill. Environment 13(2):2–12.

Blumer, M., M. Erhardt, and J. H. Jones. 1973. The environmental fate of stranded crude oil. Deep Sea Res. 20:239–259.

Bohon, R., and W. F. Clausen. 1951. The solubility of aromatic hydrocarbons in water. J. Am. Chem. Soc. 73:1571–1578.

Borneff, J., and R. Fischer. 1962. Cancerogene substanzen in wasser und boden, Mitt. X: Untersuchunger von phytoplankton eines binnensees auf polycyclische aromatische kohlenwasserstoffe. Arch. Hyg. 146:5.

Bowen, V. T. 1971. A study program to identify problems related to oceanic environmental quality. Prog. Rep. to NSF-IDOE (GX-25334), December 10, 1971. National Science Foundation, Washington, D.C. 30 pp.

Boyland, E., and J. B. Solomon. 1955. Metabolism of polycyclic compounds. 8. Acid-labile precursors of naphthalene produced as metabolites of naphthalene. Biochem. J. 59:518–522.

Brown, R. A., T. D. Searl, J. J. Elliott, B. G. Phillips, D. E. Brandon, and P. H. Monaghan. 1973. Distribution of heavy hydrocarbons in some Atlantic Ocean waters, pp. 505–519. *In* Proceedings, Joint Conference on Prevention and Control of Oil Spills. American Petroleum Institute, Washington, D.C.

Browning, E. 1953. Toxicity of industrial organic solvents. Medical Research Council, Industrial Health Research Board Rep. No. 80, rev. ed. H. M. Stationery Office, London.

Brunnock, J. V., D. F. Duckworth, and G. G. Stevens. 1968. Analysis of beach pollutants. J. Inst. Pet. 54:310–325.

Burns, K. A., and J. M. Teal. 1971. Hydrocarbon incorporation into the salt marsh ecosystem from the West Falmouth oil spill. Woods Hole Oceanographic Tech. Rep. No. 71–69. Woods Hole, Mass. 14 pp.

Burns, K. A., and J. M. Teal. 1973. Hydrocarbons in the pelagic *Sargassum* community. Deep Sea Res. 20:207–211.

Burton, M., and J. Magee. 1969. Advances in Radiation Chemistry. Wiley, New York.

Butler, J. N., B. F. Morris, and J. Sass. 1973. Pelagic tar from Bermuda and the Sargasso Sea. Spec. Publ. No. 10. Bermuda Biological Station for Research, St. George's West. AD 769 873.

Button, D. K. 1971. Petroleum-biological effects in the marine environment, pp. 421–429. *In* D. W. Hood, ed. Impingement of Man on the Oceans. Wiley-Interscience, New York.

Cahnmann, H. H., and M. Kuratsune. 1957. Determination of polycyclic aromatic hydrocarbons in oysters collected in polluted water. Anal. Chem. 29:1312–1317.

Cavanaugh, L. A., C. F. Schadt, and E. Robinson. 1969. Atmospheric hydrocarbon and CO measurements at Pt. Barrow, Alaska. Environ. Sci. Tech. 3:251–257.

Clark, R. C., and M. Blumer. 1967. Distribution of *n*-paraffins in marine organisms and sediments. Limnol. Oceanogr. 12:79–87.

Clark, R. C., Jr., and J. S. Finley. 1973a. Paraffin hydrocarbon patterns in petroleum-polluted mussels. Unpublished manuscript.

Clark, R. C., Jr., and J. S. Finley. 1973b. Petroleum hydrocarbon uptake and loss in the mussel *Mytilus edulis*. Unpublished manuscript.

Clark, R. C., Jr., J. S. Finley, B. G. Patten, D. F. Stefani, and E. E. DeNike. 1973a. Interagency investigations of a persistent oil spill on the Washington coast, pp. 793–808. *In* Proceedings, Joint Conference on the Prevention and Control of Oil Spills. American Petroleum Institute, Washington, D.C.

Clark, R. C., Jr., G. G. Gibson, and J. S. Finley. 1973b. Acute effects of outboard motor effluent on two marine shellfish. Unpublished manuscript.

Conover, R. J. 1971. Some relations between zooplankton and Bunker C oil in Chedabucto Bay following the wreck of the tanker *Arrow*. J. Fish. Res. Board Can. 28:1327–1330.

Daly, J. W., D. M. Jerina, and B. Witkop. 1972. Arene oxides and the NIH shift. The metabolism, toxicity and carcinogenicity of aromatic compounds. Experientia 28:1129–1149.

Davis, J. B. 1967. Petroleum Microbiology. Elsevier, Amsterdam. 604 pp.

Davis, W. W., M. E. Krahl, and G. H. A. Clowes. 1942. Solubility of carcinogenic and related hydrocarbons in water. J. Am. Chem. Soc. 64:108–110.

Dennis, J. V. 1959. Oil pollution survey of the United States Atlantic Coast with special reference to Southeast Florida Coast conditions. Am. Pet. Inst. Publ. 4054, Washington, D.C. 81 pp.

Diamond, L., and H. F. Clark. 1970. Comparative studies on the interaction of benzo[a]pyrene with cells derived from poikilothermic and homothermic vertebrates. I. Metabolism of benzopyrene. J. Natl. Cancer Inst. 45:1005–1011.

Dodd, E. N. 1971. The effects of natural factors on the movement, dispersal, and destruction of oil at sea. U.K. Ministry of Defense (Navy Department). (Available from NTIS, Springfield, Va., A D 763042).

Drapeau, G. 1970. Reconnaisance of survey of oil pollution on the south shore of Chedabucto Bay, March 24–25, 1970. Atlantic Oceanographic Laboratory, Dartmouth, Nova Scotia. Unpublished report.

Duce, R. A., J. G. Quinn, C. E. Olney, S. R. Piotrowicz, B. J. Ray, and T. L. Wade. 1972. Enrichment of heavy metals and organic compounds in the surface microlayer of Narragansett Bay, Rhode Island. Science 176:161–163.

Eriksson, E. 1959. The yearly circulation of chloride and sulfur in nature: Meteorological, geochemical, and pedological implications. Part 1. Tellus 11:375–403.

Falk, H. L., P. Kotin, S. S. Lee, and A. Nathan. 1962. Intermediary metabolism of benzo(a)pyrene in the rat. J. Natl. Cancer Inst. 28:699–745.

Farrington, J. W., and J. G. Quinn. 1973. Petroleum hydrocarbons in Narragansett Bay. I. Survey of hydrocarbons in sediments and clams (*Mercenaria mercenaria*). Estuarine Coastal Mar. Sci. 1:71–79.

Farrington, J. W., and G. C. Madeiros. 1973. Hydrocarbons in the surface sediments in the Western North Atlantic. Unpublished manuscript.

Farrington, J. W., J. M. Teal, J. B. Quinn, T. Wade, and K. Burns. 1973. Intercalibration of analyses of recently biosynthesized hydrocarbons and petroleum hydrocarbons in marine lipids. Bull. Environ. Contam. Toxicol. 10:129–136.

Fay, J. A. 1969. The spread of oil slicks on a calm sea, pp. 53–63. *In* D. P. Hoult, ed. Oil on the Sea. Plenum Press, New York.

Fay, J. A. 1971. Physical processes in the spread of oil on a water surface, pp. 463–468. *In* Proceedings, Joint Conference on Prevention and Control of Oil Spills. American Petroleum Institute, Washington, D.C.

Feldman, M. H. 1970a. Trace materials in wastes disposed to coastal waters: Ecological guidance and control. Federal Water Pollution Control Administration. Pacific Northwest Water Laboratory Working Paper 78. Seattle, Wash.

Feldman, M. H. 1970b. The 50-mile ballast-oil dumping prohibited zone off Alaska, reconsidered in the light of available data gleaned from significant incidents. Pacific Northwest Water Laboratory Working Paper 77. Seattle, Wash.

Filby, R. H., and K. R. Shah. 1971. Mode of occurrence of trace elements in petroleum. *In* J. R. Bogt, ed. Nuclear methods in environmental research. University of Missouri, Columbia.

Finnerty, W. R., R. S. Kennedy, B. O. Spurlock, and R. A. Young. 1973. Microbes and petroleum: Perspectives and implications, pp. 105–126. *In* Ahearn and Meyers, eds., *op. cit.*

Floodgate, G. D. 1972. Biodegradation of hydrocarbons in the sea, pp. 153–172. *In* R. Mitchell, ed. Water Pollution Microbiology. Wiley-Interscience, New York.

Floodgate, G. D. 1973. A threnody concerning the biodegradation of oil in natural waters, pp. 17–24. *In* Ahearn and Meyers, eds., *op. cit.*

Forrester, W. D. 1971. Distribution of suspended particles following the wreck of the tanker *Arrow*. J. Mar. Res. 29: 151–170.

Foster, M., M. Neushul, and R. Zingmark. 1971. The Santa Barbara oil spill. Part 2: Initial effects on intertidal and kelp bed organisms. Environ. Pollut. 2:115–134.

Freegarde, M., and C. G. Hatchett. 1970. The ultimate fate of crude oil at sea. Interim Rep. Admiralty Materials Laboratory, U.K.

Friede, J., P. Guire, R. K. Gholson, E. Gaudy, and A. F. Gaudy. 1972. Assessment of biodegradation potential for controlling oil spills on the high seas. Project Rep. No. 4110.1/3.1. Department of Transportation, U.S. Coast Guard, Office of Research and Development, Washington, D.C. 130 pp.

Gardner, L. R. 1973. The effect of hydrologic factors on the pore water chemistry of intertidal marsh sediments. Marine chemistry of intertidal marsh sediments. Mar. Chem. Unpublished manuscript.

Garrett, W. D. 1968. The influence of monomolecular surface films on the production of condensation nuclei from bubbled sea water. J. Geophys. Res. 73:5145–5150.

Garrett, W. D. 1969. Confinement and control of oil pollution on water with monomolecular surface films, pp. 257–262. *In* Proceedings, Joint Conference on Prevention and Control of Oil Spills. American Petroleum Institute, Washington, D.C.

Gebelein, C. D. 1971. Sedimentology and ecology of a carbonate facies mosaic. Ph.D. thesis. Brown University, Providence, R.I.

Girotti, R. 1968. Problems of sea pollution in the Mediterranean and fuel supply and storage in European waters. Conference on oil pollution of the sea. U.N., FAO, Rome.

Guard, H. E., and A. B. Cobet, 1973. The fate of a bunker fuel in beach sand, pp. 827–834. *In* Proceedings, Joint Conference on Prevention and Control of Oil Spills. American Petroleum Institute, Washington, D.C.

Guinn, V. P., and S. C. Bellanca. 1970. Neutron activation analysis identification of the source of oil pollution of waterways. Spec. Publ. 321, Vol. 1, p. 185 ff. National Bureau of Standards, Gaithersberg, Md.

Gunkel, W. 1967. Experimentell—ökolgische untersuchungen uber die limitierenden faktoren des microbiel en ölabbaues im marinin milieu. Helgoländer wiss. Meeresunters. 15:210–225.

Gunkel, W. 1968. Bacteriological investigations of oil-polluted sediments from the Cornish coast following the *Torrey Canyon* disaster, pp. 151–158. *In* E. B. Cowell, ed. The Biological Effects of Oil Pollution on Littoral Communities. Suppl. to Field Studies, No. 2. Institute of Petroleum, London.

Gunkel, W. 1973. Distribution and abundance of oil-oxidizing bacteria in the North Sea, pp. 127–140. *In* Ahearn and Meyers, eds., *op. cit.*

Harned, H. S., and B. B. Owen. 1958. The Physical Chemistry of Electrolytic Solutions, pp. 531–534. Reinhold, New York.

Harvey, G. R., and J. M. Teal. 1970. PCB and hydrocarbon contamination of plankton by nets. Bull. Environ. Contam. Toxicol. 9:287–290.

Heyerdahl, T. 1971. Atlantic Ocean pollution and biota observed by the "Ra" expedition. Biol. Conserv. 3:164–167.

Holcomb, R. W. 1969. Oil in the ecosystem. Science 166:204–206.

Hollinger, J. P., and R. A. Mennella. 1973. Measurement of the distribution and volume of sea surface oils using multi-frequency radiometry. Science 181:54–56.

Horn, M. H., J. M. Teal, and R. H. Backus. 1970. Petroleum lumps on the surface of the sea. Science 168:245–246.

Hoult, D. P. 1972. Oil spreading on the sea. Annu. Rev. Fluid Mech. 4:341–368.

Jeffrey, L. M. 1973. Preliminary report on floating tar balls in the Gulf of Mexico and Caribbean Sea. Sea Grant Project 53399, Texas A&M University. Unpublished report.

Jeffrey, P. G. 1973. Large-scale experiments on the spreading of oil at sea and its disappearance by natural factors, pp. 469–474. *In* Proceedings, Joint Conference on Prevention and Control of Oil Spills. American Petroleum Institute, Washington, D.C.

Johnson, B. H., and T. Aczel. 1967. Analysis of complex mixtures of aromatic compounds by high-resolution mass spectrometry at low ionizing voltages. Anal. Chem. 39:682–685.

Johnson, R. 1970. The decomposition of crude oil residues in sand columns. J. Mar. Biol. Assoc. 50:925–937.

Karczewska, H. 1972. Microbial degradation of hydrocarbons in mineral oils—A literature survey. Rep. B-136. Swedish Water and Air Pollution Research Laboratory, Stockholm. 48 pp.

Kator, H. 1973. Utilization of crude oil hydrocarbons by mixed cultures of marine bacteria, pp. 47–66. *In* Ahearn and Meyers, eds., *op. cit.*

Keizer, P. D., and D. C. Gordon. 1973. Detection of trace amounts of oil seawater by fluorescence spectroscopy. J. Fish. Res. Board Can. 30:1039–1046.

Kinney, P. J. Hydrocarbon biodegradation in Port Valdez. Institute of Marine Sciences Report: Port Valdez Environmental Studies. In press.

Klug, M. J., and A. J. Markovetz. 1971. Utilization of aliphatic hydrocarbons by microorganisms, pp. 1–43. *In* A. H. Rose and J. F. Wilkinson, eds. Advances in Microbial Physiology, Vol. 6. Academic Press, New York.

Kolpack, R. L., J. S. Mattson, H. B. Mark, and T. C. Yu. 1971. Hydrocarbon content of Santa Barbara Channel sediments, Vol. II, pp. 276–295. *In* Biological and Oceanographical Survey of Santa Barbara Channel Oil Spill, 1969–1970. Allan Hancock Foundation, University of Southern California. Los Angeles.

Kreider, R. E. 1971. Identification of oil leaks and spills, pp. 119–124. *In* Proceedings, Joint Conference on Prevention and Control of Oil Spills. American Petroleum Institute, Washington, D.C.

Lee, R. A. 1973. Uptake of petroleum hydrocarbons by marine copepods. Unpublished manuscript.

Lee, R. A., and A. A. Benson. 1973. Fate of petroleum in the sea—Biological aspects, pp. 541–551. *In* Background papers, workshop on petroleum in the marine environment. National Academy of Sciences, Washington, D.C.

Lee, R. F., R. Sauerheber, and A. A. Benson. 1972a. Petroleum hydrocarbons: Uptake and discharge by the marine mussel, *Mytilus edulis*. Science 177:344–346.

Lee, R. F., Sauerheber, and G. H. Dobbs. 1972b. Uptake, metabolism and discharge of polycyclic aromatic hydrocarbons by marine fish. Mar. Biol. 17:201–208.

Levy, E. M. 1972. Evidence for the recovery of the waters off the coast of Nova Scotia from the effects of a major oil spill. Water Air Soil Pollut. 1:144–148.

Lisitzin, A. P. 1972. Sedimentation in the world ocean. Soc. Econ. Paleontol. Miner. Spec. Publ. 17. 218 pp.

Liu, D. L. S. 1973. Microbial degradation of crude oil and the various hydrocarbon derivatives, pp. 95–104. *In* Ahearn and Meyers, ed., *op. cit.*

Ludwig, H. F., and R. Carter. 1961. Analytical characteristics of oil–tar materials on Southern California beaches. J. Water Pollut. Control Fed. 33:1123–1139.

McAuliffe, C. 1966. Solubility in water of paraffin, cycloparaffin, olefin, acetylene, cycloolefin and aromatic hydrocarbons. J. Phys. Chem. 70:1267–1273.

McAuliffe, C. 1969a. Solubility in water of normal C_9 and C_{10} alkane hydrocarbons. Science 158:478–479.

McAuliffe, C. 1969b. Determination of dissolved hydrocarbons in subsurface brines. Chem. Geol. 4:225–233.

McCarthy, R. D. 1964. Mammalian metabolism of straight-chain saturated hydrocarbons. Biochem. Biophys. Acta 84:74–79.

MacIntyre, F. 1970. Geochemical fractionation during mass transfer from sea to air by breaking bubbles. Tellus 22:451–462.

Mallet, L., and J. Sardou. 1964. Examination of the presence of the polybenzic hydrocarbon benzo-3,4-pyrene in the planktonic environment of the Bay of Villefranche. Symposium, Committee on International Scientific Exploration of the Mediterranean Sea, Monaco. C. R. Acad. Sci., Paris 258:5264–5267.

Meinschein, W. G. 1969. Hydrocarbons—Saturated, unsaturated, and aromatic. *In* G. Eglinton and M. T. J. Murphy, eds. Organic Geochemistry. Springer, New York.

Meyers, P. A. 1972. Association of fatty acids and hydrocarbons with mineral particles in seawater. Ph.D. thesis, University of Rhode Island, Providence.

Miget, R. J. 1973. Bacterial seeding to enhance biodegradation of oil slicks, pp. 291–309. *In* Ahearn and Meyers, eds., *op. cit.*

Mitchell, R., S. Togel, and I. Chet. 1972. Bacterial chemoreception: An important ecological phenomenon inhibited by hydrocarbons. Water Res. 6:1137–1140.

Mize, C. E., J. Avigan, D. Steinberg, R. C. Pittman, H. M. Pales, and G. W. A. Milne. 1969. A major pathway for the mammalian oxidative degradation of phytanic acid. Biochem. Biophys. Acta 176:720–739.

Monaghan, P. H., and C. B. Koons. 1973. Petroleum in the marine environment: Gulf of Mexico. Gulf Coast Association of Geological Societies Proceedings, Houston, Tex.

Monaghan, P. H., J. H. Seelinger, and R. A. Brown. 1973. The persistent hydrocarbon content of the sea along certain tanker routes. A preliminary report, API tanker conference, Hilton Head Island, S.C., May 7–9.

Morris, B. F. 1971. Petroleum: Tar quantities floating in the Northwestern Atlantic taken with a new quantitative neuston net. Science 173:430–432.

Morris, B. F., and J. N. Butler. 1973. Petroleum residues in the Sargasso Sea and on Bermuda beaches, pp. 521–530. In Proceedings, Joint Conference on Prevention and Control of Oil Spills. American Petroleum Institute, Washington, D.C.

Moulder, D. S., and A. Varley. 1971. A bibliography on marine and estuarine oil pollution. The laboratory of the Marine Biological Association of the United Kingdom, Plymouth, Devon, U.K.

Noshkin, V. E., and V. T. Bowen. 1973. Concentration and distributions of long-lived fallout radionuclides in open ocean sediments, pp. 671–686. In Radioactive Contamination of the Marine Environment. International Atomic Energy Agency, Vienna.

Noyes, W. A., and P. A. Leighton. 1966. Photochemistry of Gases. Dover Publ. Co., New York.

Parker, C. A. 1970. The ultimate fate of crude oil at sea—uptake of oil by zooplankton. AML Rep. B. 198(M).

Parker, C. A. 1971. The effect of some chemical and biological factors on the degradation of crude oil at sea, pp. 237–244. In P. Hepple, ed. Water Pollution by Oil. Institute of Petroleum, London.

Parker, P. L., J. K. Winters, and J. Morgan. 1972. A base-line study of petroleum in the Gulf of Mexico, pp. 555–581. In Base-line studies of pollutants in the marine environment (heavy metals, halogenated hydrocarbons and petroleum). National Science Foundation, IDOE, Washington, D.C.

Pasteels, J. J. 1968. Pinocytose et athrocytos par l'epithelium branchial de Mytilus edulis. Z. Zellforsch. 92:239–259.

Paterson, M. P., and K. T. Spillane. 1969. Surface films and the production of sea-salt aerosol. Q. J. R. Meterol. Soc. 95:526–534.

Peake, E., and G. W. Hodgson. 1966. Alkanes in aqueous systems. I. Exploratory investigations on the accommodations of C_{20}–C_{33} n-alkanes in distilled water and occurrence in natural water systems. J. Am. Oil Chem. Soc. 43:215–222.

Peake, E., and G. W. Hodgson. 1967. Alkanes in aqueous systems. II. The accommodation of C_{12}–C_{36} n-alkanes in distilled water. J. Am. Oil Chem. Soc. 44:696–702.

Polikarpov, G. G., V. N. Yegorov, V. N. Ivanov, A. V. Tokareva, and I. A. Feleppov 1971. Oil areas as an ecological niche. Priroda 11 [transl. by N. Precoda], Pollut. Abstr. 3:72–5TC–0451.

Programmes Analysis Unit. 1973. The environmental and financial consequences of oil pollution from ships. Appendix 2. The fate of oil at sea. Report of Study No. VI, submitted to the Intergovernmental Maritime Consultative Organization by the United Kingdom. 10 pp.

Quinn, J. G., and T. L. Wade. 1972. Lipid measurements in the marine atmosphere and sea surface microlayer, pp. 633–664. Background paper for Brookhaven (24–26 May 1972) workshop on baseline studies of pollutants in the marine environment. In E. D. Goldberg, ed. International Decade of Ocean Exploration. National Science Foundation. Washington, D.C.

Ramsdale, S. J., and E. E. Wilkinson. 1968. Identification of petroleum sources of beach pollution by gas-liquid chromatography. J. Inst. Pet. 54:326–332.

Revelle, R., E. Wenk, B. H. Ketchum, and E. R. Corino. 1971. Ocean pollution by petroleum hydrocarbons, pp. 297–318. In W. H. Matthews, F. E. Smith, and E. D. Goldberg, eds. Man's Impact on Terrestrial and Oceanic Ecosystems. MIT Press, Cambridge, Mass.

Risebrough, R. W. 1969. Chlorinated hydrocarbons in marine ecosystems, pp. 5–23. In M. W. Miller and G. C. Berg, eds. Chemical Fallout: First Rochester Conference on Toxicity. Charles C. Thomas, Springfield.

Risebrough, R. W., P. Reiche, D. B. Peakall, S. G. Herman, and M. N. Kirven. 1968. Polychlorinated biphenyls in the global ecosystem. Nature 220:1098–1102.

Robertson, B., S. Arhelger, P. J. Kinney, and D. K. Button. 1973. Hydrocarbon biodegradation in Alaskan waters, pp. 171–184. In Ahearn and Meyers, eds., op. cit.

Rosenberg, E., and D. Gutnick. 1973. Bacterial growth and dispersion of crude oil in an oil tanker during the ballast voyage. Unpublished manuscript.

Roubal, W. T. 1973. In vivo and in vitro spin labeling studies of pollutant-host interaction. In F. J. Biros and R. Haque, eds. Proceedings, Symposium on Mass Spectrometry and NMR of Pesticides. Plenum Press, New York.

Scarratt, D. J., and V. Zitko. 1972. Bunker C oil in sediments and benthic animals from shallow depths in Chedabucto Bay, Nova Scotia. J. Fish. Res. Board Can. 29:1347–1350.

Seidell, A. 1941. Solubilities of Organic Compounds. D. Van Nostrand, New York.

Seifert, W. K., and W. G. Howells. 1969. Interfacially active acids in a California crude oil; isolation of carboxylic acids and phenols. Anal. Chem. 41:554–562.

Sever, J. R., T. F. Lytle, and P. Haug. 1972. Lipid geochemistry of a Mississippi coastal bog environment. Contrib. Mar. Sci. 16:149–161.

Sherman, K., J. B. Colton, R. L. Dryfoos, B. S. Kinnear. 1973. Oil and plastics contamination and fish larvae in surface waters of the Northeast Atlantic. MARMAP Operational Test Survey Report: July–August 1972, January–March 1973. Unpublished manuscript.

Shipton, J., J. H. Last, K. E. Murray, and G. L. Vale. 1970. Studies on a kerosene-like taint in mullet (Mugil cephalus). II. Chemical nature of the volatile constituents. J. Sci. Food Agric. 21:433–436.

Simonov, A., and A. Justchak. 1970. The effect on the chemical content of sea water with a limited exchange of water from a large ocean of polluting discharges of chemical (with the Baltic Sea as an example). Advances in Water Pollution Research. Proceedings of the Fifth International Conference, San Francisco and Hawaii, Vol. 2. W. H. Jenkins, ed. Pergamon Press, London.

Simpson, A. C. 1968. The Torrey Canyon disaster and fisheries. Ministry of Agriculture, Fisheries and Food, Fisheries Laboratory, Burnham on Crouch, U.K. laboratory leaflet (new series) 18. 43 pp.

Smith, J. E. 1968. Torrey Canyon Pollution and Marine Life. Cambridge University Press, London.

Sohngen, J. L. 1913. Benzin, petroleum, parafinöl und paraffin als kohlnestoff und energiequelle für mikroben. Zentralbl. Bakt. Parasit. Infekt. II 37:596–609.

Spooner, M. 1969. Some ecological effects of marine oil pollution, pp. 313–316. Proceedings, Joint Conference on Prevention of Control of Oil Spills. American Petroleum Institute, Washington, D.C.

Spooner, M. 1970. Oil spill in Tarut Bay, Saudi Arabia. Mar. Pollut. Bull. 1:166–167.

Stegeman, J. J., and J. M. Teal. 1973. Accumulation, release, and retention of petroleum hydrocarbons by the oyster *Crassostrea virginica*. Mar. Biol. 22:37–44.

Suess, E. 1968. Calcium carbonate interactions with organic compounds. Ph.D. thesis, Lehigh University, Bethlehem, Pa.

Suess, E. 1972. Laboratory experimentation with 3, 4 benzpyrene in aqueous systems and the environmental consequences. Aus Wissenschaft und Praxis. Zentralbl. Bakt. Hyg. I. Abt. Orig B 155:541–546.

Tissier, M., and J. L. Oudin. 1973. Characteristics of naturally occurring and pollutant hydrocarbons in marine sediments, pp. 205–214. *In* Proceedings, Joint Conference on Prevention and Control of Oil Spills. American Petroleum Institute, Washington, D.C.

Traxler, R. W. 1973. Bacterial degradation of petroleum materials in low temperature marine environments, pp. 163–170. *In* Ahearn and Meyers, eds., *op. cit.*

Vale, G. L., G. S. Sidhu, W. A. Montgomery, and A. R. Johnson. 1970. Studies on a kerosene-like taint in mullet (*Mugil cephalus*). I. General nature of the taint. J. Sci. Food Agric. 21:429–432.

Voroshilova, A. A., and E. V. Dianova. 1950. Bacterial oxidation of petroleum and its fate in natural waters. Mikrobiologiya 19:203–210.

Woodwell, G. M., C. F. Wurster, and P. A. Isaacson. 1967. DDT residues in an East Coast estuary: A case of biological concentration of a persistent insecticide. Science 156:821–824.

Yevich, P. P., and M. M. Barry. 1970. Histopathologic finding in molluscs exposed to pollutants. Abstract of papers, session III, section 4, 33d annual meeting, American Society of Limnology and Oceanography, University of Rhode Island, August 25–29.

Zafiriou, D. C. 1972. Response of *Asterias vulgaris* to chemical stimuli. Mar. Biol. 17:100–107.

Zafiriou, D. C. 1973. Petroleum hydrocarbons in Narragansett Bay. II. Chemical and isotopic analysis. Estuarine Coastal Mar. Sci. 1:81–87.

Zitko, V. 1971. Determination of residual fuel oil contamination by aquatic animals. Bull. Environ. Control Toxicol. 5:559–564.

ZoBell, C. E. 1964. The occurrence, effects, and fate of oil polluting the sea, pp. 85–109. *In* Proceedings, International Conference on Water Pollution Research, London, 1962. Pergamon Press, London.

ZoBell, C. E. 1969. Microbial modification of crude oil in the sea, pp. 317–326. Proceedings, Joint Conference on the Prevention and Control of Oil Spills. American Petroleum Institute, Washington, D.C.

ZoBell, C. E. 1971. Sources and biodegradation of carcinogenic hydrocarbons, pp. 441–451. *In* Proceedings, Joint Conference on Prevention and Control of Oil Spills. American Petroleum Institute, Washington, D.C.

ZoBell, C. E. 1973. Microbial degradation of oil: Present status, problems, and perspectives, pp. 3–16, 153–162. *In* Ahearn and Meyers, eds., *op. cit.*

Zsolnay, A. 1972. Preliminary study of the dissolved hydrocarbons and hydrocarbons on particulate material in the Götland Deep of the Baltic. Kieler Meeresforsch. 27:129–134.

Zsolnay, A. 1973a. The relative distribution of non-aromatic hydrocarbons in the Baltic in September 1971. Mar. Chem. 1:127–136.

Zsolnay, A. 1973b. Hydrocarbon and chlorophyll: a correlation in the upwelling region off West Africa. Deep Sea Res. 20:923–925.

4 Effects

The effects of oil spills may be acute or chronic in nature. Acute effects on the biota are those that result from a single infusion of oil into the marine environment from an accidental spill. Mortalities due to petroleum and its by-products may occur almost simultaneously with, or at any time after, the appearance of the oil in the environment. The effects may be due to chemical or physical characteristics of the petroleum. One such effect is the smothering or asphyxiation of organisms by a coating of oil.

These accidental spills constitute a small fraction (3 to 4 percent) of the annual rate of addition of petroleum into the marine environment. Some of these spills occur within confined marine areas, such as bays or estuaries, where the concentration may remain high for extended periods causing the biological impacts to be greater than if the oil were released where rapid dispersion could take place. Such releases are generally large compared with chronic low-level additions and, furthermore, they commonly occur in coastal waters where man makes maximum use of marine resources. A list of some of the more important oil spills is given in Table 4-1.

The public is concerned when seabirds at nearby beaches and nesting areas are oiled and when shellfish are killed or tainted such that they are unpalatable. When beaches are polluted with oil, public use decreases.

Chronic effects are those that occur from the release of crude oil or its derivatives either continuously or sufficiently often that the biota does not have time to recover between doses. Other activities associated with the petroleum and other industries may have a long-term or chronic effect on aquatic organisms. For example, dredging and filling of marshes can modify organism habitats and either kill or have sublethal effects on individual organisms. The sublethal modifications may affect the characteristics of the populations of each species, changing the rates of birth, death, and dispersal, as well as the age structure and spatial pattern. Also, changes in the ecological communities may occur in the affected area. A list of various sublethal effects on the physiology, histology, and behavior of organisms and on their populations are described in Table 4-2.

There are various levels of biological effects of oil. At various places in the marine environment and at various times these will be accorded different priorities in the evaluation of the impact. These effects include the possibility of:

1. Human hazard through eating contaminated seafood;
2. Decrease of fisheries resources or damage to wildlife such as seabirds and marine mammals;
3. Decrease of aesthetic values due to unsightly slicks or oiled beaches;
4. Modification of the marine ecosystem by elimination of species with an initial decrease in diversity and productivity;
5. Modification of habitats, delaying or preventing recolonization.

In this report, an exhaustive bibliography has not been included because such bibliographies are already available (for example, United Kingdom Report No. 6

TABLE 4-1 A Summary of Several Major Oil Spills Followed by Studies of Their Biological Impact

Date of Spill	Source and Location	Type and Amount of Oil (barrels)	Shoreline Affected (mi)	Localities Studied	Species Identified	Sampling Method	Biological Damage	Reference
March 1957	*Tampico Maru*, Baja California, Mexico	Diesel oil 60,000	2	Intertidal and subtidal	Larger visible plants and animals	Qualitative, quantitative macrocystis counts	Nearly total devastation immediately, luxuriant growth of seaweed developed within months; biota 90% restored after 3 or 4 years, although relative abundance of certain species still somewhat changed after 12 years	North et al., 1964; Mitchell et al., 1970
July 1962	*Argea Prima*, Guayanilla Harbor, Puerto Rico	Crude oil 70,000		Mangrove shores; intertidal and subtidal	Blue-green algae	Qualitative	Extensive damage: high mortalities among many shallow water and shore-dwelling organisms, including a wide variety of vertebrates; also extensive damage to intertidal and sublittoral algae and mangrove habitat	Diaz-Piferrer, 1962
Jan. 1967	*Chryssi P. Goulandris*, Milford Haven, England	Crude oil >1,800		Intertidal salt marsh; intertidal rocky shore	Grasses	Semiquantitative rocky shore transect; quantitative studies of grasses	Most damage to intertidal organisms; gastropod molluscs badly affected, also barnacles and sea anemones on a number of shores; no apparent damage to algae	Cowell, 1969; Nelson-Smith, 1968
March 1967	*Torrey Canyon*, S.W. England	860,000[a]		Intertidal rocky shores and sand beach	Larger visible animals only	Semiquantitative rocky shore transects; qualitative beach and subtidal surveys; quantitative algal counts	Very high mortalities of intertidal shore life, mostly due to use of toxic emulsifiers; many invertebrates and algae killed on shores; fisheries and plankton apparently unaffected; estimated 10,000 birds killed	Bellamy et al., 1967; Smith, 1968
Sept. 1967	*R. C. Stoner*, Wake Island	Aviation gas, J-P4 jet fuel, A-1 turbine oil, and Bunker C oil 126,000		Intertidal and subtidal	Large visible invertebrates	Qualitative	Many dead fish stranded on shores; also abundant dead molluscs, sea-urchins, and crabs	Gooding, 1968
March 1968	*Ocean Eagle*, San Juan Harbor, Puerto Rico	Crude oil 83,000		Intertidal rocky shore	15 large sp.	Qualitative	Many subtidal and intertidal organisms killed or damaged by oil or oil and emulsifier, including molluscs, crustaceans, and algae, although subsequent recovery good; 10 species of fish found dead or in state of stress	Cerame-Vivas, 1968
April 1968	*Esso Essen*, S. Africa	Crude oil 20,000–28,000		Intertidal and subtidal	No species identifications, observations on larger organisms	Qualitative	High mortalities of sandhoppers (amphipods) but otherwise little damage on shores; high bird mortalities	Stander and Ventner, 1968

74

Date	Spill	Amount (tons)	Habitat	Species studied	Method	Effects	Reference
Dec. 1968	*Witwater*, Galeta Island, Canal Zone	Diesel and Bunker C oil 20,000	Rocky intertidal, coral reef, sandy intertidal mangroves	*Uca*, mangrove species, four coral species	One quantitative sand sample for meiofauna; otherwise qualitative	On rocky shores, extensive mortality of supralittoral vegetation and tide pool life; on sandy beaches, great population decreases among meiofauna, especially crustaceans; many young mangroves killed in swamp areas, also algae and many invertebrates; coral reefs apparently unharmed	Rutzler and Sterrer, 1970
Jan. 1969	Well A-21, Santa Barbara Channel	Crude oil 33,000[a]	Intertidal and subtidal	Subtidally: selected polychaete families, ophiuroids, and molluscs not including smaller polychaetes and amphipods; intertidally: visible rocky shore species and 195 sp. retained by 1.5-mm screens in sandy areas	40 Grab sample, qualitative at species level; quantitative for biomass line transects on rocky shores; 1/100 m³ samples on beaches	High mortalities of intertidal organisms covered with oil; about 3,600 birds killed; no apparent effects on fish and plankton; no directly attributable damaging effects of oil on large marine mammals or on benthic fauna; area recovering well within a year	Cimberg et al., 1973; Fauchald, 1971; Foster et al., 1971a,b; Nicholson and Cimberg, 1971; Straughan, 1972
Sept. 1969	*Florida*, West Falmouth, Mass.	No. 2 fuel oil 4,500[a]	Intertidal mud and sand flats; subtidal to 10 mm	All animals 0.247 mm, excluding nematodes, copepods; ostracods and unicellular organisms, including smaller polychaetes and amphipods	3 Quantitative transects	Severe pollution of sublittoral zone, with 95% kill of all fauna, including many fish, worms, molluscs, crabs, lobsters, and other crustaceans and invertebrates; local shellfish industry severely affected; Wild Harbor still closed to shellfish fishing in May 1974	Blumer and Sass, 1972; Blumer et al., 1970a,b
Feb. 1970	*Arrow*, Chedabucto Bay	Bunker C 108,000[a]	Intertidal rocky shore; intertidal lagoon	Common visible species on rocky shore and species 74 mm in lagoon samples	12 Semiquantitative transects; 2 samples in lagoon	Localized damage to intertidal life, where most mortalities were crabs, limpets, and algae, probably killed by smothering; local fish catches normal; about 2,300 birds killed; 5 months after spill, subtidal flora and fauna healthy; fishing and lobstering normal	Thomas, 1973; Navships, 1970
Jan. 1971	*Arizona Standard* and *Oregon Standard*, San Francisco Bay	Bunker C 20,000[a]	Intertidal and subtidal rocky shore; intertidal sand beach	31 larger sp.	60 Quantitative transect counts	Some damage to shore life, mainly to acorn barnacles, limpets, mussels, and striped shore crabs; 3,600 birds killed; area nearly normal within 1 year	Chan, 1973
Feb. 1971	*Wafra*, Cape Aulhas, S. Africa	Crude oil 445,000	Intertidal rocky shores	Larger intertidal rocky shore species	10 Qualitative	Little damage to intertidal life; 1,135 black footed penguins found oiled	Day et al., 1971
April 1971	March Point Dock Facility, Anacortes, Washington	No. 2 fuel oil 5,000	Intertidal beaches, rocky shores, subtidal	Animals 4 mm in subtidal samples, visible epifauna in intertidal areas; fauna identified to major taxa only	20 Quantitative grabs; quantitative intertidal transects	Some oil on shores, damaging shellfish, limpets, crabs, clams and oysters; about 1,000 birds involved	Watson et al., 1971; Woodin et al., 1973
Jan. 1972	*General M.C. Meigs*, Wreck Cove, Washington Coast	Navy special oil 3,000[a]	Intertidal rocky shores	37 sp. algae, sp. animals not including smaller polychaetes and amphipods	300–500 yd Quantitative transects	Urchins affected; plant community showed less of fronds and bleached thalli	Clark et al., 1973

[a]Moore, Stephen F. et al. 1974. Potential biological effects of hypothetical oil discharges in the Atlantic coast and Gulf of Alaska. *In* Report to Council on Environmental Quality, MIT Sea Grant Prog. MITSG 74-19, Cambridge.

TABLE 4-2 Evaluation of Experiments and Observations of the Sublethal Effects on Organisms Both of Pollution and of Other Associated Activities of the Petroleum Industry

Group	Species	Reference	Type of Petroleum Product	Concentration	Effects and Evaluation
A. Reproduction					
1. Fecundity					
Crustacea	*Pollicipes polymerus*	Straughan, 1971	Crude oil, Santa Barbara blowout field study		Inverse relationship between the fraction of adults brooding and the amount of oil on the adults (p 0.5); heavily and moderately oiled areas had no recruitment whereas settlement was recorded from all unoiled samples
Mollusca	*Mytilus edulis*	Blumer et al., 1971	No. 2 fuel oil, West Falmouth spill field observations		Gonads of mussels failed to develop in affected areas
2. Fertilization and Development					
Marsh grass	*Festuca rubra*	Baker, 1971	Kuwait ("fresh," 10% distillate and 90% residue)	Seeds soaked in 5 ml of various oils for 1 h	Germination almost completely inhibited; experiments limited in scope, but related to field situation
Polychaeta	*Capitella capitata*	Bellan et al., 1972	Detergent ("low" toxicity polyethylene-glycol fatty acid)	0.01–10 ppm	Survival of young decreased; reduction in number of females laying eggs; most complete study of sublethal effects on reproduction; laboratory strain of this hardy species probably reflects extreme tolerance conditions
Fish	*Gadus morhua*	Kuhnhold, 1970	Iranian crude extracts (paraffin based)	Aqueous extracts from 10^4, 10^3, 10^2 ppm total oil (author estimates 10^4 yields 10 ppm soluble hydrocarbons, 1 ppm may be more likely)	*Eggs:* "Some cases" were sublethal but embryos and larvae did not survive, apparently, 10^2 ppm does not differ from control. *Larvae.* "Showed typical behavior symptoms in oil extracts: increased activity was followed by a reduction of swimming activity, which finally stopped ... which slowly deepened until the 'critical point' when no responses of the larvae were obtained even by touching or prodding"; time to "critical point" varies with age of larvae and amount of oil: 10^2 ppm not different from control (14–5.5 days for 1–10-day-old larvae); 10^3 ppm (8.4–4.5); 10^4 ppm (4.2–0.5); "herring larvae were less, and plaice larvae more, resistant than cod"; "chemoreceptors seemed to be blocked very quickly at the first contact with oil"; insufficient quantification; no measure of uncertainty; no chemical analysis
	Pleuronectes platessa	Wilson, 1970	BP1002	0–10 ppm	*Larvae:* "1/50 of 100 C_{50} (0.04–0.2 ppm BP1002) disrupt photoactic and feeding behavior (the ability of the larvae to capture prey items was significantly impaired)"; recovery in uncontaminated seawater in 1–3 days); ranges for control in photoactic response experiments not given; feeding results not documented
Lobster	*Homarus americanus*	Wells, 1972	Venezuelan crude	0.1, 1, 6, 10, and 100 (emulsions)	100 ppm lethal to all larval stages; 10 ppm; stages 1–3 more sensitive than stage 4. *Long-term experiments with newly hatched larvae:* 10 ppm: 9-day mean survival time; 6 ppm: longer time to 4th, longer than at lower concentrations; concentrations at which development was prolonged are too high to be important in the field

B. Growth

Phytoplankton	Several diatoms and dinoflagellates	Mironov, 1970	Concentrations of oil mixed with seawater (1.0–0.001 ml/l); effective concentration *may be* 0.1 ppm–0.1 ppb	No cell division or delayed cell division, compared with controls; dinoflagellates generally more susceptible than diatoms; no mention of type of oil in this work; no measure of concentration of soluble hydrocarbons; concentrations reported seem unrealistically low
	Unspecified microalgae, *Asterionella japonica*	Aubert et al., 1969	Kerosene	3 ppm; 38 ppm	Depression of growth rate; concentration of soluble hydrocarbons unknown
	Phaeodactylum tricornutum	Lacaze, 1967; Nelson-Smith, 1973	Kuwait crude	1% extracts, effective concentration may be 1 ppm	Depression of growth rate; concentration of soluble hydrocarbons unknown
	Chlorella vulgaris	Kauss et al., 1973	Aqueous extracts of several crude oils and outboard motor oil; 90% solutions of aqueous extracts used	1 part oil to 20 parts water	Inhibition of growth varied from 5 to 41% after 2 days of exposure; after 10 days, cell yields were close to controls, suggesting inhibiting substance was eventually lost. After 2 days of cell growth, cell numbers were significantly lower in 25, 50, and 90% oil extracts than in control; concentrations of water-soluble hydrocarbons and comparison of oils unknown
	Monochrysis lutheri	Strand et al., 1971	Kuwait crude; dispersant (holl. chem. 622) emulsions	25–500 ppm solutions of benzene; 25–250 ppm solutions of toluene; 100 ppm solutions of xylene; 3–27 ppm naphthalene 20–100 ppm	Inhibition of growth of 1–2 days' duration (volatization eventually reduced effect); minimum concentrations tested had an effect but are unrealistically high; lower concentrations should have been tested Inhibition of growth during 6 days of experiment at all concentrations tested; lowest concentrations should have an effect, therefore, lower concentrations should have been studied; effect of crude-dispersant emulsion difficult to relate to other information
	Phaesdactylum tricornutum, Skeletonema costatum, Chlorella sp., *Chlamydomonas* sp.	Nuzzi, 1973	Extracts of outboard motor oil, No. 6 fuel oil, No. 2 fuel oil	1 ppm	Inhibition of growth during 10–12 days incubation with No. 2 fuel oil only at 20% of extracted medium, stimulation of growth by crude and motor oil extracts; concentration of water-soluble hydrocarbons not determined, therefore, it is difficult to compare oils
Plants	*Distichlis maritima, Festuca rubra*	Baker, 1971	Kuwait atmospheric residues		Treatment Apply: 4/23/70 Apply: 7/11/69 Harv.: 6/10/70 Harv.: 8/27/69 Unoiled 5.4 ± 0.8 2.9 ± 0.4 4 l/m² 7.8 ± 1.4 2.9 ± 0.7 8 l/m² 11.7 ± 0.7 1.6 ± 0.5 *Results*: increase in dry weight (g/25 cm quadrat); similar for *F. rubra* but stimulation not observed with *Spartina anglica*; experiments with "fresh" Kuwait reduce growth; numerous additional experiments reported; results not reported systematically, statistical significance included
Mollusca	*Crassostrea virginica*	Mackin and Hopkins, 1961	Locally produced crude	Spray	Sprayed directly on oysters weekly for 6 months; no effect on growth or survival of adult oysters or spat
			Heavy loss of crude by wild well; Port Sulfur, La.; started 1/17/56, oil not evident by 3/72		No effect on growth, nor any difference in growth between experiment and controls for all size oysters

TABLE 4-2 (Continued)

Group	Species	Reference	Type of Petroleum Product	Concentration	Effects and Evaluation
B. Growth (Continued) Mollusca (Continued)	*Crassostrea virginica*	Mackin and Hopkins, 1961	Bleedwater		Oysters in trays and bags at varying distances from outfall; *Effect:* mortality at 50–75 ft; reduced growth and glycogen content at 75–150 ft; none beyond 150 ft
	Littorina littorea	Perkins, 1970	BP1002	30 ppm	Significant inhibition of growth
C. Metabolism 1. Photosynthesis Phytoplankton	Mixed natural samples	Gordon and Prouse, 1973	Venezuelan crude No. 2 and No. 6 fuel oil	10–200 µg/l (ppb)	Concentrations below 10–30 µg/l were found to stimulate photosynthesis, while at concentrations between 60 and 200 µg/l, were somewhat suppressed below controls for all but No. 2 fuel oil which depressed photosynthesis to approximately 60% of controls at concentrations between 100 and 200 µg/l; environment in Bedford Basin: 0.5–60 µg/l; highest content (under slick): 800 µg/l
	Monochrisis lutheri	Strand et al., 1971	Kuwait crude; dispersant (holl. chem. 622) emulsions	1–1,000 ppm	Significant reduction of bicarbonate uptake at concentrations 50 ppm; effect of crude-dispersant emulsion difficult to relate to other information; only observations recorded; no data presented in this preliminary report
	Chlamydomonas angulosa	Kauss et al., 1973	Naphthalene	3–24 ppm	Almost complete reduction of bicarbonate uptake at concentrations as low as 3 ppm in stoppered flasks; when volitization permitted uptake rebounds; lowest concentration tested produced reduction; therefore, experiments at lower, probably more realistic, concentrations should have been conducted
Kelp	*Marcocystis pyrifera*	North, 1965			
Lichen	*Lichen pygmaea*	Brown, 1972	Kuwait crude BP1002	0.1–100 ppm	Total ^{14}C fixation was affected by emulsifier at concentrations of 1 ppm and greater; 50% fixation below control at 100 ppm and near 0 fixation at 50% emulsifier–at higher concentrations tested, there is also an increase in carbohydrate leakage to the seawater from the lichen—also conducted experiments to test effects of aging oil and/or emulsifier but results gave either no difference or they were unclear; oil and emulsifier gave same response as emulsifier and seawater, so oil was not the important toxicant; cites field studies which confirm lab results in that emulsifier and not oil kills lichens; concludes that since aging didn't increase the toxicity of PB1002, then it is the surfactant and not the volatile solvent that is of significance in toxicity
2. Respiration Fish	*Cyprinodon variegatus, Lagodon thomboides, Micropogon undulatus*	Steel and Copeland, 1967	Petrochemical wastes	0.2–2.0 ppm in addition 0.4–4.0 phenol	Claims respiratory inhibition at low concentration, then stimulation approaching TL 48 as general pattern; only 1 of the 3 species fit this pattern; insufficient acclimation; too high concentrations

	Lagodon rhomboides	Wohlschlag and Cameron, 1967	Petrochemical wastes	50% of wastewater	Depression of respiration much more severe at near-lethal temperatures than intermediate ones; water collected from Corpus Christi turning basin then diluted to ½; so achievable in the field; lack of confidence intervals
Molluscs	*Mytilus edulis, Modiolus demissus*	Gilfillan, 1973	Extract of midcontinent sweet crudes	1% emulsion	Differential effects of salinity and oil on filtering rate and respiration; normal salinity: increasing oil produces erratic stimulation of respiration; depression of filtering rate above 1% extract; 21 ppt salinity increases respiration and increases filtering rate up to 10% extract, then a very rapid dropoff at higher concentrations of oil. 11 ppt: shutdown of activity; no additional effect of oil
Fish	Juvenile *Onchorhynchus tshanytsha* (salmon) and *Morone saxatilis* (striped bass)	Brocksen and Bailey, 1973	Benzene	5 and 10 ppm changed every 48 h	Respiratory rate was increased during the early (24–48-h) period of exposure to both 5 and 10 ppm of benzene; after longer periods, respiration decreases back to near-control levels; when tested at daily intervals after exposure, found both fish species returned to control levels
D. Behavior					
Fish	*Ictalurus natalis*	Todd, 1972			Feeding unaffected; social behavior altered after 1–3 days and returned to normal in about 1 week; second additions after return to normal again disrupted social behavior
	Notemigonus crysoleucas				No apparent effect on social behavior and/or feeding
Crustacea	*Homarus americanus*	Atema and Stein, 1972	La Rosa crude and extracts thereof		Change in feeding times (doubling of waiting time) and behavior caused by addition of 1:100,000 parts crude oil to water; soluble fractions giving same oil/water ratio had no effect; light and electron microscopy showed no change in morphology of odor receptors; response similar over 5-day period, although hydrocarbons' characteristics did change by "weathering"; experimentally good as possible, but long-term effects and recovery not considered; very difficult problem
Gastropoda	*Nassarius obsoletus*	Jacobson and Boylan, 1973	Kerosene extract	1–4 ppb	Significant reduction in attraction to oyster extract (0.31 ppm); significant reduction in attraction to scallop extract (3 ppm); analytically and statistically sound; experimental concentrations realistic to chronically polluted harbors and spill-affected areas
Crustacea	*Pachygrapsus crassipes*	Kittredge, 1971	Crude oil (Sisquoc sand in California) dilutions (as low as 1:100) of diethyl ether extracts		Inhibition of chemically mediated feeding and 6 pheromone responses in the presence of dilutions of crude oil extracts; response recovery in 3–6 days; quantification less than adequate, not statistically tested; method of extract preparation of questionable validity; concentration of hydrocarbons not known
	Uca pugnax	Krebs, 1973	No. 2 fuel oil, West Falmouth spill, heavily and moderately oiled marshes		Results were that male and females exhibited breeding display colors, and males exhibited threat posture even though the breeding season had been over; mortality was heavy in these areas; interpretations based on observational information only; little quantification and adequate control areas lacking

TABLE 4-2 (Continued)

Group	Species	Reference	Type of Petroleum Product	Concentration	Effects and Evaluation
D. Behavior (Continued)					
Fish	*Onchorhynchus gorbuschka* (salmon)	Rice, 1973	Prudhoe Bay crude oil		Small fry are much more tolerant of oil and avoid it at much higher concentrations than large fry; large fry: TL$_m$ 48–110 ppm, avoidance at 1.6 ppm. The speculation that this avoidance could disrupt migrations may be reasonable since such an effect has been reported for zinc pollution on Atlantic salmon
E. Histological Changes					
Fish	*Menidia menidia*	Gardner, 1972	Texas-Louisiana crude oil	1 l w/40 l seawater, then separate fractions (soluble and insoluble); exposed for 168 h	Various types of histological abnormalities exhibited after exposure to both the soluble and insoluble fractions; largely chemoreceptor structures studies but also ventricular myocardium; no analyses of concentrations seen by fish; technique seems good, but tissue and water content of hydrocarbons needed
Clam Fish	*Mya arenaria* *Menidia menidia*	Gardner et al., 1973	Texas-Louisiana crude oil and No. 2 fuel	600 ppm	0.95 l crude oil mixed with 37.85 l seawater; after settling, oil on top (water-insoluble fraction) at 600 ppm used to expose *Menidia*, also water layer (water-soluble fraction) at 126 ml/l used on other groups of *Menidia*; some histological damage to chemoreceptors of fish but concentrations seen by fish not known; *Mya arenaria* collected and sectioned 4 months after a spill of No. 2 fuel oil showed an increased incidence of gonadal tumors, compared with those collected from control area; all results indicative of possibly good technique but effect not well quantified and neither was content of hydrocarbons; other causes for tumors possible
Mollusca	*Crassostrea virginica*	St. Amant, 1970	Dredging, etc., associated with petroleum extraction in marshes of Louisiana coast		Activities have caused increased incursion of higher salinity water into marshes that formerly had lower salinity; decreased oyster production
Benthos Zooplankton Neuston		Mackin, 1971	Brine effluents from 5 different oil fields; volume and characteristics differ widely		Over a radius of approximately 400 ft, the effluence from Trinity Bay and Fisher's Reef fields affects the size and diversity of benthic communities; no other effect from other distances, other fields, or the zooplankton or neuston was found. A quite complete coverage of species and conducted over a 1-year period; would be good if had hydrocarbon concentration data
Benthos		Reish, 1965	Refinery effluent		Bottom of L.A. harbor that received wastes from refineries was either uninhabited or inhabited only by *Capitella capitata*; after a part of region dredged, number of species increased, then rapidly declined again; correlated increase of carbon in the sediment; region studies limited to only a mile of narrow channel near the source of the effluents; in addition, this part of the harbor received wastes from many other sources; not certain that the oil was actually responsible, nor how far and to what dilu-

80

	Reish, 1971	Refinery effluent		By 1970, all L.A. refineries ceased discharging wastes or eliminated the oxygen depleting fraction; number of benthic species increased from 0 in 1954 at the four most heavily polluted stations to 4, 6, 10, 11 in November 1970; work sketchy, but does demonstrate that oil wastes definitely were responsible for the impoverished benthic fauna before abatement; no samples from greater distances to suggest the extent of the effect by refinery wastes
Marsh	Baker, 1971	Refinery effluent hydrocarbon; characteristics not specified	About 2 ppm phenols	*S. anglica* gradually killed over 20-year period, probably by repeated coating by thin oil films; in creek blue-green algae have replaced filamentous greens (*S. anglica* will grow in soil samples and effluent water samples)
Spartina anglica community				
Marsh grasses	Baker, 1971	Successive spillage of Kuwait crude		Results, ranking from least to most tolerant species: (1) shallow rooted plants with no or small food reserves (*Suaeda maritima, Salicornia* sp.); (2) shrubby perennials (*Halimione portulacoides*); (3) filamentous green algae (*Ulothrix* sp., *Baucheria* sp.); (4) perennials (*Spartina townsendi, Juncus maritima, Puccinellia maritima, Festuca rubra, Agrostis stonifera*); (5) perennials of rosette habit with large food reserves (*Cochlearia* sp., *Glaux maritima, Artemisia maritima*)
Various	Crapp, 1971	Benthic Kuwait crude emulsifier BP1002		Communities, hard substrates, Milford Haven: (a) same shores compared after 7–9 years, only species changing were also changed elsewhere due to temperature changes over a large area; (b) field experiments crude oil had little detectable effect, but emulsifiers reduced herbivorous molluscs with the result that inner tidal alga increased; (c) stranded weathered crude, mechanical smothering of some inner tidal animals
	Crapp, 1971; Baker, 1973	Refinery effluents: phenol oils, etc.		Effluents poorly described, only "oil," "phenols," etc.; grazing snails reduced in numbers, particularly small ones, nearer (27 m) vs. further (1960 m) from outfall; fucoid algae more abundant at nearer stations; prior to this refinery, the shore had few algae, mostly barnacles and limpets; same situation in 1970, 1972
Invertebrates	Nicholson and Cimberg, 1971	Santa Barbara crude and natural seep		Coal Oil Point had a less varied invertebrate fauna than other comparable areas along the coast of Southern California; suggested that chronic exposure to the nearby natural seep might account for the low diversity by eliminating sensitive kinds; sampling along single line transects inappropriate for assessing the variety of organisms that inhabit a locality; Onuf added 21 more species to their list of 24 in 1½ h
Mussels	Kanter *et al.*, 1971	Crude	1,000, 10,000, 100,000 ppm	Coal Oil Point mussels were more resistant than ones from other areas, suggesting that chronic exposure leads to selection for tolerant forms; alternative that inherent physiological variability between populations may account for differences in oil tolerance is not eliminated and is suggested by the 10–100-fold difference in tolerance of mussels from 2 nonseep area samples
Mytilus californianus				

TABLE 4-2 (Continued)

Group	Species	Reference	Type of Petroleum Product	Concentration	Effects and Evaluation
D. Behavior (Continued)					
Fish	Gulf of Mexico sp.	Bechtel and Copeland, 1970	Petrochemical wastes	Different % Houston Ship Channel water	Claims that percent polluted water is a good predictor of species diversity; unfounded because confounded with salinity; to convincingly demonstrate that oil pollution responsible for decreased diversity, compare with samples covering a similar range of salinites in an unpolluted control bay
Fish Shrimp	Gulf of Mexico sp.	Spears, 1971	Oil field wastes	~ 18 ppm	Yields of harvestable fisheries products less in every case in the one creek receiving oil field wastes than in 5 nearby relatively undisturbed creeks: 4–30 times fewer shrimp, 2–25 times fewer blue crabs, 5–16 times fewer game fish, 1.4–11 times fewer forage fish; adverse effects may extend to the bay receiving the oil filled waters; on average, half as many stations on state tracts yielded harvestable organisms in the bay receiving oil field wastes; study useful because it does incorporate reasonable controls; drawbacks in applying to other situations: the brine is at least as toxic as the soil, control bay is not so variable in salinity, and rate of dilution in the receiving bay is undetermined

for Inter-Governmental Maritime Consultative Organization, 1973; Moulder and Varley, 1972; Boesch, 1973). Only critical references for conclusions are quoted. Knowledge and understanding of the effects of oil in the marine environment are by no means complete and are in many cases inadequate. Critical unanswered questions are identified, and studies should be encouraged to resolve the problems these pose.

FACTORS INFLUENCING THE BIOLOGICAL IMPACT OF OIL SPILLS

When an oil spill occurs, many factors determine whether that spill will cause heavy, long-lasting biological damage, comparatively little or no damage, or some intermediate degree of damage. An example of the variability that exists among the effects of oil spills on the marine biota is outlined by Mitchell et al. (1970) in his description of the widely different effects resulting from the *Tampico Maru* and the Santa Barbara oil spills. Several factors apparently contributing to the difference in those effects are compared. Straughan (1972) also identifies various factors that influence the extent of biological damage from an oil spill.

OIL DOSAGE

If the spill occurs in a small, confined area so that the oil is unable to escape, damage will be greater, almost without exception, for a given volume and type of oil spilled than if that same volume were released in a relatively open area. At the *Tampico Maru* site, 60,000 barrels of diesel fuel were spilled in a small cove, whereas at Santa Barbara 75,000 barrels of crude oil were spilled in an open ocean channel. The *Tampico Maru* spill had greater, more persistent damage than the Santa Barbara spill, most likely because it occurred in a confined area. Thus, where the spill occurs, as well as the volume of oil spilled, are important factors.

OIL TYPE

The *Tampico Maru* spill involved No. 2 diesel fuel and the Santa Barbara spill involved crude oil (Mitchell et al., 1970); because No. 2 diesel fuel oil contains more aromatic compounds, it is more toxic. The type of oil spilled may be the most important factor, as evidenced by the difference in severity accompanying the spill by the barge *Florida* at West Falmouth, Massachusetts (Blumer et al., 1971), which was carrying No. 2 diesel fuel oil, and the *Torrey Canyon* spill, which involved Kuwait crude oil (Smith, 1968). A definitive study of the difference in toxicity among various oils is reported by Ottway (1970), who showed that the susceptibility of a snail, *Littorina littorea,* varied markedly according to which of 20 test oils it was exposed.

OCEANOGRAPHIC CONDITIONS

Currents, wave action, and coastal formation all combine to influence the dosage of a given spill. Currents and wave action—especially in an open body of water such as a large bay, channel, or the open ocean—dilute the spilled oil, thereby reducing the toxicity of the spilled oil (Straughan, 1972).

METEOROLOGICAL CONDITIONS

Normally, storms will increase wave action and thereby reduce the toxic properties of the spilled oil by dilution. On occasion, however, wave action may intensify the problems, as apparently occurred at the West Falmouth spill. Soon after this spill, the surf drove the oil ashore into the sediments and the surrounding marshland. The oiled marshland and sediments then became a reservoir of oil for many months (Blumer et al., 1971).

TURBIDITY

The amount of turbidity in the water is significant in determining the fate of the spilled oil. If the water contains suspended sediments or solids, much of the oil will sink. At Santa Barbara (Kolpack, 1971) the spill occurred during a period of heavy storms that caused considerable freshwater runoff. Sediments that were introduced by the flood waters into the nearby coastal waters provided an adsorptive surface for the spilled oil. Upon wetting the sediment surfaces, the oil flocculated the particles and subsequently sank. Such sedimentation is advantageous if the intertidal life is abundant, but it may be detrimental to benthic life (Blumer et al., 1971).

SEASON

The season of the year in relation to the life cycle of a given organism can frequently be the factor determining whether the effect of a spill is severe or relatively light (Straughan, 1972). For example, if a spill occurs during the winter in an area where seabirds are nesting, bird mortality may be in the thousands; at some other time of the year, the mortality would be much lower. Similarly, if a spill occurs in an estuary when salmon fry are going to sea or during a salmon run, a much higher kill is likely (Anonymous, 1955). Crab larvae, which float near the surface of the water in their zoea form, will probably be killed if a spill occurs during this stage of their life cycle (Mironov, 1969). Newly set oyster spats are also exceedingly vulnerable. However, such damage would not occur at other times of the year when these organisms are at a different stage of development.

BIOTA TYPE

Straughan (1972) reports that at Santa Barbara the barnacle *Balanus glandula* was not as easily smothered by oil as the barnacle *Chthamalus fissus*. *Balanus glandula* is larger in size and can survive a thicker layer of oil encrusted on the shoreline. Also, it has a calcareous base plate, whereas the smaller barnacles do not; because the plate prevented direct exposure to the substrate, it resettled earlier. According to Straughan's observations, the type of biota is important in assessing damage.

METHODS OF OIL SPILL CLEANUP

At present, mechanical methods are the least damaging methods of cleaning up oil spills (Straughan, 1972; Gatteleir et al., 1973). Such methods involve the use of booms and skimmers or absorbents. Retrieval of spilled oil by booms and skimmers is not effective if the current exceeds 1.7–1.0 knots, if the wave height is greater than 1–2 f, or if the oil is a distillate product. Absorbents also have limited application. Proper techniques and equipment for evenly distributing large quantities of sorbents over wide areas of open water, for properly agitating and mixing the sorbents with the oil mass, for harvesting the oily agglomerate, and for processing or disposing of the recovered oil-absorbent mass are not available (Struzewski and Dewling, 1969).

Sinking agents, such as sand and stearated chalk, are another method of cleanup. These materials were used by the French to sink large masses of oil that spilled from the *Torrey Canyon* incident in the Bay of Biscay (Struzewski and Dewling, 1969) and have been developed further in Holland (Meijs et al., 1969). Although no adverse effects on fisheries and benthic life were reported, the lack of knowledge on the precise fate of the oil would indicate that further experiments are needed before this method can be recommended. Sinking may extend the period that the benthic fauna may be affected (Struzewski and Dewling, 1969).

The use of dispersants is another controversial method. According to Cowell (1971) and Smith (1968), most of the damage that occurred at the *Torrey Canyon* spill was caused, not by the use of dispersants, but by their misuse. Specifically, the dispersants were applied undiluted to oil after it had come ashore. Moreover, Beynon (1970) and Canevari (1971) point out that since the *Torrey Canyon* incident, dispersants have been developed that are far less toxic and, if properly used, pose a minimum threat to or burden on the marine environment.

Those opposed to the use of dispersants claim that dispersing the oil into the water column renders the oil easier for marine organisms to assimilate (Blumer, 1969a,b; Murphy and McCarthy, 1970). Dewling (1971) pointed out that the use of dispersants, especially in rivers and estuaries, imposes an added burden on the assimilative capacity of the river or estuarine system to biodegrade the oil/dispersant mixture.

Straughan (1972) points out that in certain circumstances such as the protection of an endangered species of birds, the use of low-toxicity dispersants may override all other considerations. Beynon (1971) stresses the advantage of using dispersants on oil spills to prevent the oil from washing ashore and killing intertidal organisms because of its toxicity or by smothering. The application of the dispersant must be while the oil slick is still far enough from shore so that the concentration of the oil/dispersant mixture is quickly diluted below its toxic level (Beynon, 1971). Further, several spills that used low-toxicity dispersants were subsequently surveyed and showed no apparent biological damage (Canevari, 1973).

Gatteleir et al. (1973) reports that for a number of years in France dispersants have been used routinely with minimal environmental effect. The advantage of using dispersants is the dilution effect that reduces toxicity and facilitates biodegradation. Straughan (1972) expresses doubt as to the effectiveness of many of these dispersant products in the open sea if inadequate mixing occurs. However, Canevari (1973) reports the recent development of low-toxicity dispersants that do not require the use of mixing energy to effect dispersal of the oil. The dispersant has a driving force to diffuse across the oil water interface and, in effect, achieves spontaneous dispersion of the oil

Because opinion is polarized concerning the use of dispersant, research under field conditions is needed to establish the conditions and circumstances under which dispersants can be used effectively.

EFFECTS ON VARIOUS TYPES OF HABITATS

INTERTIDAL AREAS

Intertidal organisms contend with alternating periods of immersion and exposure, sudden infusions of fresh water, greatly elevated salinities induced by evaporation, freezing, radiant heating, and other stresses. Organisms in this fluctuating habitat are much hardier than the subtidal forms living in a more constant environment. Exposure of stranded oil to the aerial environment accelerates evaporation of the shorter chain hydrocarbons and thus reduces toxicity. With loss of these shorter chain components, oil can become tarry and coat intertidal surfaces. As a result, stress from surface coating is increased in intertidal areas and may affect some organisms.

ROCKY SUBSTRATES

Oil trapped in tide pools on falling tides can saturate small volumes of water with dissolved organics for periods ranging to several hours. Toxicities might become intensified as a result. Oil deposits on attached epifauna and flora can result in layers several centimeters thick. Usually such material is at least partially removed by the rising tide. Toxic effects may appear where stranded oil contacts tissues directly, such as on plant life (Holmes, 1969). Smothering may develop when tarry residues build up (Straughan, 1971).

SANDY SUBSTRATES

Oil stranded on beaches on falling tides tends to percolate into the sediments, which provides an opportunity for intimate contact with the infauna. Sedimentary transport processes can move the oily particles away from the region of initial contact, thus extending them along the shore as well as to subtidal waters.

Any oil absorbed by sediment grains may be exposed when the wave scour removes the overburden. Logically, oil dispersed as fine films on small particles should be ideally suited for microbial attack, but rapid degradation requires abundant oxygen, and sediments may easily become anerobic, thus delaying degradation by microorganisms. Oil trapped in sandy English beaches persisted for several months (Smith, 1968) and was not released until later by various agents.

MARSHLAND

Oil stranded in marshland also enters sediments and can penetrate at least 70 cm (Burns and Teal, 1971). Oil residues were detected in organisms from various levels of the food web approximately a year after the West Falmouth spill in the Wild Harbor River Marsh (Burns and Teal, 1971). Also, substantial quantities of residues were recovered about 1½ years after the West Falmouth spill.

Estuaries and marshlands, huge areas on the east coasts of Virginia, South Carolina, Florida, and the Gulf, often serve as spawning and nursery areas. Studies of the estuarine and marshy waters affected by the West Falmouth spill indicate that oil spilled in this type of habitat could have far-reaching effects if spawning and nursery functions became affected. Blumer *et al.* (1970a) recorded an unexpected rise in sedimentary oil content of the estuarine muds in Wild Harbor many months after the initial spill. This rise was attributed to a release of oily material from the nearby marshland. The new infusion caused additional adverse effects among the infauna (Sanders *et al.*, 1972). The system was thus able to "store" oil for later release.

Salt marsh plants were killed by a refinery effluent released in sheltered tidal creeks at Southampton, England (Baker, 1971), but effluents released in more exposed water with rapid dispersion seem to have had little biological effect (Baker, 1973).

SUBTIDAL SUBSTRATES

Usually, oil floats on the sea surface, but several mechanisms can sink it, thus exposing subtidal communities. For oil to sink, sedimentary particles must contact the slick, forming a denser aggregate of oil-coated particulates that can settle. Such processes may occur in surf (North *et al.*, 1969), from stranding in the intertidal area (Blumer *et al.*, 1970a), or from contact with turbid runoff (Kolpack, 1971).

EFFECTS ON AQUATIC ORGANISMS

Some insight into the immediate and long-term effects of oil spilled in marine waters on the physiology and behavior of organisms, as well as the structure and function of their communities, can be developed by reviewing the effect of major spills that have occurred during the past few decades (Tables 4-1 and 4-2). The most extensive tabulation of major oil spills, with an attempt to assess the environmental impact of each spill, is reported by Ottway (1972).

In most of the major spills that occurred during the time period between 1960 and 1971 (Ottway, 1972), the environmental impact assessment was made subjectively; that is, without the benefit of trained personnel using scientific methods. However, more than a dozen spills have been studied by trained personnel in varying degrees of thoroughness (Table 4-1).

The resulting biological damage, as determined by these studies, ranged from comparatively light damage, as exemplified by the spills from the *Esso Essen* (Stander and Ventner, 1968) and the *Wafra* (Day et al., 1971, to nearly total immediate devastation of marine organisms, as shown by the *Tampico Maru* (Mitchell et al., 1970) and the *Florida* (Blumer and Sass, 1972a) spills.

Oil may be introduced with brines that are associated with its production. Separation of petroleum from aqueous phases is never complete. When the brines are discharged into coastal water, they may carry with them from 10 to 50 ppm oil and even higher quantities of other noxious substances. The soluble fractions of petroleum probably are the most harmful to marine organisms. These undoubtedly make up the bulk of the oil that enters the coastal waters as co-product brines and refinery effluents. Such discharges, particularly when they occur in estuaries, may pollute the shallow water areas that serve as nursery areas for many coastal marine biota. These waters ordinarily are subject to extremes of temperature, and where tital fluctuations are small, flushing of the bays and dispersal of wastes may be slow. These characteristics of estuaries may combine with the chronic introduction of oil wastes from brines to intensify biological effects.

In addition, the slow dispersal of oil in some estuaries can increase adsorption onto suspended particulate matter. When these suspended particulates are incorporated in the sediments, the exposure of benthic organisms to oil is increased. The high biological productivity of estuaries ensures the exposure of many organisms to the wastes, particularly in the sensitive stages of their life history. Synergistic interactions between extremes of temperature and salinity and oil may aggravate deleterious effects. Therefore, because they combine biological productivity with the most severe exposure to wastes, estuaries are most vulnerable to the serious effects of chronic oil pollution.

PRODUCTIVITY

Biological effects from large oil spills vary so widely that it is difficult to generalize on any specific topic such as productivity. Damage to plants can occur during acute phases of spills. Such effects have been recorded for phytoplankton (Mironov, 1971), attached algae (North et al., 1969), and flowering plants (Holmes, 1969). Documentation that shows plant life surviving virtually unharmed after oil spills is also available (Smith, 1968; Straughan, 1972). Plants are relatively resistant to toxicity from oil and require substantial exposure under natural conditions before significant damage results.

Productivity may be limited by the availability of nutrients or by activities of herbivores. Nutrient availability is probably not affected by oil, but destruction of herbivores can occur. Many macroinvertebrate grazers require months or years to develop. If grazing populations are seriously reduced by catastrophe, algal blooms of rapidly growing species can appear. Thus, a population explosion of the rapidly growing brown alga *Macrocystis* was observed after the wreck of the tanker *Tampico Maru* (North et al., 1969). The appearance of the kelp bed was attributed to lack of control by missing herbivores, primarily sea urchins.

Presumably similar effects may occur in a planktonic situation. The time scale involved in phytoplankton blooms due to similar causes is much shorter than for kelp beds and other attached algae.

PRIMARY PRODUCERS

Studies have been conducted concerning the capacity of marine plant species, including phytoplankton, kelp, and a lichen, to fix ^{14}C-labeled bicarbonate during exposure to a variety of petroleum hydrocarbons and/or oil dispersants. Responses to hydrocarbons ranged from no reduction in photosynthesis by lichens coated with Kuwait crude (Brown, 1972) to a 40 percent reduction in carbon fixation exhibited by a mixed natural population of phytoplankton subjected to 100–200 ppb of No. 2 fuel oil (Gordon and Prouse, in press). No. 6 fuel oil stimulated photosynthesis at concentrations below 10–30 ppb, although at concentrations between 60 and 200 ppb photosynthesis was repressed. However, Gordon (personal communication) reports that there is no effect on growth rate for certain species up to concentration of 300 ppb of either crude or No. 2 fuel oil. In laboratory tests with a specific hydrocarbon, the lowest concentration of napthalene tested (3 ppm) caused almost complete elimination of bicarbonate uptake by *Chlamydomonas angulosa* (Kauss et al., 1973). Other effects of various types of oil on phytoplankton are described in Table 4-1.

Clendenning reported that emulsions of fuel and diesel oils in seawater strongly inhibited photosynthesis by large kelp under laboratory conditions (California State Water Pollution Control Board, 1964).

DETRITUS FEEDERS

The West Falmouth studies provide certain insights into possible effects of an oil spill on productivity. In this specific case, the term "productivity" is strictly defined as the amount of living protoplasm produced during a given interval of time. Numbers can be substituted for the *productivity measurement* if the discussion is confined to the intertidal flats along Wild Harbor River or restricted shallow-water bottoms where the initial kill was almost total. These nearby azoic sediments were very quickly invaded by the polychaete worm *Capitella capitata,* which was one of the few species that survived, grew, and reproduced in the oil contaminated sediments of Los Angeles Harbor, where a total of 1.5 million barrels of oil industry wastes were discharged per day in 1957 (Reish, 1965).

In September 1969, samples were taken from Wild Harbor River to measure *Capitella* abundance (Sanders *et al.,* 1972). These samples showed that *Capitella* abundance was approximately 2,700/m^2. Within 3 weeks after settling onto the bottom sediments, the larvae of this opportunistic species become sexually mature. This characteristic, together with a remarkably broad tolerance to stress, a short generation time, and huge reproductive potential, enabled *Capitella,* in the absence of other species in any density, to increase exponentially in numbers at these intertidal stations. By December 1969, a population of more than 200,000 *Capitella*/m^2 was achieved, converting the intertidal flats into a dense unispecies cluture of *Capitella.*

Because *Capitella* is a deposit-feeder, the killoff and decay of the resident populations of the flora and fauna, together with additions from the affected neighboring marshes in September of 1969, provided a highly nutrient-rich substrate for it. The extremely high productivity of *Capitella* was made possible by the absence of competitors and predators and the lack of food or space restrictions.

Obviously, such an increase could not continue for any extended period of time, for the polycheate became ultimately space- or food-limited even in the absence of other species. The precipitous drop in number of *Capitella* during July, August, and September 1970 to about 1,000 individuals per square meter coincided with the reappearance of other deposit-feeding species and predators (Sanders *et al.,* 1972).

REPRODUCTION

Very little data have been reported on effects of petroleum substances on fecundity of marine organisms. However, some field experiments indicate that seed germination can be almost completely inhibited by coating with Kuwait crude oil (Baker, 1971). Other studies show that the hatching of eggs of several species of birds have been inhibited because they were coated with several types of oil by the oiled feathers of their parents Hartung, 1965).

GROWTH

The experiment on marsh grasses is the one most relevant to effects on growth (Baker, 1971). Growth may be stimulated or reduced, depending on the type of substance applied, season of application, and frequency of application. The marsh grasses *Puccinella maritima* and *Festuca rubra* increased in dry weight production in the presence of atmospheric residues of Kuwait crude (Baker, 1971). However, *Distichlis spicata* was killed by repeated application of crude oil during a 5-month period (Mackin, 1970).

Oysters are highly resistant to repeated or continuous exposure to petroleum substances (Mackin and Hopkins, 1961). Spraying oysters with crude oil for 6 months did not affect their growth or survival. Coproduct brine killed oysters at a range of 50–75 ft from an outfall. At the same range, growth and glycogen content were reduced; however, no effect was reported beyond 150 ft.

RESPIRATION

Initial investigations regarding the effects of petrochemical wastes on the respiratory response of estuarine fish demonstrated some change in oxygen consumption (Wohlschlag and Cameron, 1967). However, because of the short period of measurement, lack of acclimation, and absence of detailed water analysis, the work was not judged conclusive. Further, the waste water used contained numerous substances, including acids and phenol, which were probably more toxic than the petroleum hydrocarbons. It was suggested that near-lethal temperatures in combination with chemical pollutants act synergistically to modify normal respiratory rates (Wohlschlag and Cameron, 1967). A similar synergistic effect was indicated by Gilfillan (1973) in a study on the effects of salinity and an extract of crude oil on *Mytilus edulis* feeding and respiration. Although it was not stated whether some emulsification occurred during extraction, the oil extract caused a decrease in feeding and an increase in respiration. Measurements of these same parameters at salinities below full strength seawater (21 and 11 ppt of salinity) indicated that they were responding to the combined stresses of salinity and oil and then to salinity stress alone at 11 ppt.

The effect of benzene on the respiratory rates of juvenile salmon and striped bass have been examined (Brocksen and Bailey, 1973). In these tests, the water was analyzed for benzene content and then differences in the weight of fish were corrected by regression analysis. The oxygen consumption of both species generally increased after exposures of 24–48 h at concentrations of 5 or 10 ppm. After being placed in clean water for approximately 6 days, the fish returned to control levels of respiration.

BEHAVIOR

The behavioral responses of various marine organisms

to low (ppb range) concentrations of petroleum hydrocarbons have been studied (Todd, 1972). The high complexity of petroleum substances and the complexity of behavioral responses, such as feeding and reproduction, produce a very difficult problem to analyze and to investigate. Some experiments have been performed but were preliminary (Todd, 1972). The complex chemical changes that occurred in the experiments (Atema and Stein, 1972) hindered effective interpretation of the results. However, results do indicate that lobster feeding behavior may be affected by the emulsified whole oil rather than by soluble hydrocarbons. The effects of kerosene extracts (ranging from 1 to 4 ppb soluble hydrocarbons) were investigated on the attraction of the snail *Nassarius obsoletus* to oyster and scallop extracts (Jacobson and Boylan, 1973).

HISTOLOGY

The tissue damage resulting from exposure of *Menidia menidia* and *Mya arenaria* to oil and other toxic materials have been examined (Gardner, 1972; Gardner et al., 1973). Although the volumes of oil and water used were described, the actual concentrations in the water were not analyzed. Deterioration of the chemoreceptor structures of *Menidia* was found after exposure to both soluble and insoluble fractions of Texas–Louisiana crude oil for 168 h; *Mya* collected from an area subjected to a spill of No. 2 fuel oil had a higher incidence of gonadal tumors than those collected from a control area (Gardner et al., 1973).

EVALUATION OF AVAILABLE INFORMATION

When the literature pertaining to the sublethal effects of petroleum hydrocarbons on the marine organisms was reviewed, several common shortcomings were identified. The most notable among these is the absence of competent quantitative and qualitative analyses of control and experimental water characteristics. Three studies reviewed included analyses of hydrocarbons present in the exposure medium: Gordon and Prouse (in press), Atema and Stein (1972), and Jacobson and Boylan (1973). Techniques are available that permit a detailed analysis of the petroleum derived hydrocarbons in animal tissues and seawater samples under certain conditions (Anderson, 1973).

Effects on Aquatic Populations and Communities

Oil at very low concentrations, which causes no direct mortality, may have some adverse effects on populations and, in turn, on natural communities. Because organisms have different tolerances to oil and oil products, changes in relative abundances can occur. Within a trophic level, resistant species may flourish when populations of sensitive species decline and make available previously limited resources. Between trophic levels, a reduction in the number of grazers can lead to drastic habitat changes with the establishment of a canopy of macroalgae (North et al., 1969) and the shading out of the diatom flora on the substrate. Subtle disturbances to populations and communities could arise from or because of interference with chemical communications involved in feeding and mating.

PLANKTON

Studies on phytoplankton (Hohn, 1959) and zooplankton (Odum et al., 1963) of the Galveston Bay, Texas, system indicate decreased species diversity in the area near the Houston ship channel. This area is heavily burdened with petrochemical and other toxic wastes. The effects of lowered salinity and other toxicants compound the picture. Field evidence regarding the effects of chronic oil pollution on planktonic communities is lacking.

SUBTIDAL ORGANISMS

For benthic organisms, more studies are conducted for accidental spills than for chronic inputs. Refinery effluents may have a considerable impact on the benthic life in confined bodies of water where dispersion of the effluent is not rapid (Baker, 1973). For example, animals inhabiting sediments in Los Angeles harbor, which received large quantities of oil and other industrial wastes, were eliminated or limited to a single tolerant polychaete, *Capitella capitata* (Reish, 1965). The greatest effects were caused by the depletion of oxygen on the bottom by oxygen-demanding wastes that concentrated in the sediments.

Oil-burdened sediments can be toxic to rocky bottom and sedentary organisms (North et al., 1969; Sanders et al., 1972; Burns and Teal, 1971). Subtidal organisms, such as kelp, that extend throughout the water column can trap oil in their surface canopies. During the Santa Barbara spill, although kelp beds collected large quantities of oil and tar, no harm was observed (Jones et al., 1969). Kelp blades are usually slimy with mucoid substances. This mucoid substance may have prevented direct contact between plant tissues and oil. Subtidal communities attached to Platform A, the site of the Santa Barbara channel spill, were inspected about 2 weeks after the spill. Minor mortalities among pelecypods and barnacles were noted (Jones et al., 1969) when oil globules were rising slowly through the water.

Damage occurs, as in the *Torrey Canyon* incident, where thick oil films are washed ashore (Smith, 1968). In this case, immobile organisms were affected most; mortality resulted by smothering. To a much smaller degree, organisms died from the chemical impact of the oil. Fish and some very mobile crustaceans escaped because they could move away from the endangered places and, therefore, were not observed dead in appreciable numbers. Where chemical products had been used to disperse the oil and to clean the beaches, more organisms died than in nontreated areas; the chemicals used (surfactants plus organic solvents) proved toxic. Resettlement was slower in areas where chemicals had been sprayed on a large scale (Straughan, 1971).

Biological observers of the *Torrey Canyon* disaster concluded that spilling a large quantity of crude oil leads to deleterious effects of a local character. When a productive shellfish area is contaminated by an oil spill, this natural resource will be adversely affected. However, such local events will probably not lead to extermination of any given species because the geographical range of virtually all marine organisms living in shallow water is much larger than a single affected area.

In open bays the area in which benthic life is affected by refinery effluents or co-product brine discharges generally seemed to be confined to a radius of several hundred meters (Baker, 1973).

FISHERIES

Any possible effect on fishery productivity is of prime concern in coastal areas where there are chronic inputs of petroleum hydrocarbons. One of the most extensive areas of coastal petroleum development and also an area of tremendous fishery productivity is in Louisiana. More than 25,000 producing wells are situated in coastal Louisiana, with some oil fields that have been in production for more than 40 years and many that have existed for at least 20 years. The chronic addition of oil through co-product brines is probably about twice the addition caused by accidental spills. Annual additions of petroleum at the estimated rates over the past 30 years would mean that the Louisiana coastal waters have received 1.1 million barrels of oil. However, commercial fishing catches continue high in Louisiana waters (Table 4-3).

Although evidence of mortality in chronically polluted areas is rare, tainting of oysters is frequently reported and is generally associated with sediments containing high levels of petroleum hydrocarbons (500 ppm). Tainted oysters must be removed to unpolluted areas for several months to make them marketable.

Other activities of the petroleum industry in coastal Louisiana have had more damaging effects on the coastal ecosystem and fisheries than oil pollution. These activities include direct physical damage (including dredging and silting effects from the construction of canals necessary for the laying of pipelines and drilling rigs), physical displacements of marshlands and subtidal bottoms, and damage caused by improper seismic operations.

Far more serious, however, are the indirect effects of the petroleum industry activities, including accelerated erosion of wetlands, altered tidal flow patterns, and consequent greater salinity intrusion into the estuaries. The annual loss of marshland to erosion in Louisiana has been calculated at 16.5 sq mi, and the total loss over the past 30 years was put at 316,797 acres (Gagliano and van Beek, personal communication). Actual canaling has accounted for more than 13 percent of this total loss. As a result of the alteration of tidal current patterns, whole embayments have silted up in some areas while, in other areas, new ones have been formed. Salinity intrusion has resulted in a shifting inland and compression of the low salinity nursery grounds, which are important for estuarine-dependent fisheries.

Although the total yield of oysters and shrimps in coastal Louisiana waters has not decreased, definite changes have occurred. The original oyster grounds in brackish water were destroyed by the incursion of higher salinity water accompanied by predators and fungal disease. The new grounds are apparently less suitable because the yield per acre in 1972 fell to almost 10 percent of the 1945 figure (Table 4-3). The composition of the shrimp catch has shifted, from 95 percent white shrimp (*Penaeus setiferus*) to about 50 percent of this species and 50 percent brown shrimp (*P. aztecus*). The apparent reason is that the brackish water nursery grounds of *P. setiferus* in Louisiana have been reduced. Finally, the production of muskrats has declined because their food plants have diminished with decrease in the low salinity marshes.

The period that the Louisiana fisheries can withstand these alterations to the coastal environment remains unknown. That the fisheries have not seriously declined reflects the great reproductive potential and resiliency of these oyster and shrimp species. Through an overproduction of dispersive larvae, an oversupply of food, and a short generation period (1 year), they manage to sustain the continued pressure.

The effect of chronic oil pollution on the fishery resources of other coasts is also not well known. The North Sea receives oil from England and continental Europe, from frequent tanker spills and from offshore oil production facilities. However, to separate these effects from those of other pollutants and the effects of fishing itself is impossible (Korringa, 1968). The Caspian Sea

TABLE 4-3 Oyster and Shrimp Yields in Coastal Louisiana Waters

Year	Oysters Pounds (× 1,000)	Area in Acres	Number of Dredges	Shrimp Pounds[a] (× 1,000)	Number of Boats
1939	13,586	–		88,000	1,621
1940	12,412	–		98,000	
1945	9,884	19,760	500	116,904	
1950	8,715	–		77,835	2,819
1951	8,164	36,000	226	85,718	
1955	9,396	–		83,608	
1957	10,489	–		34,103	
1960	8,311	58,000	143	61,758	4,896
1961	10,139	–		31,027	
1964	11,401	–		59,382	
1965	8,343	88,500	94	62,593	7,296
1969	–			89,500	10,320
1970	8,639	112,000	77	92,600	12,500
1971	9,758	113,000	86	95,000	
1972	8,947	157,000	56	87,000	14,500
1973	–	161,200			

[a] Heads on.

receives large inputs of petroleum from production fields and refineries (Kasymov, 1971). Total fishery yield has declined from 300 million kg to 110 million kg in 35 years. [This is probably due in part to a drop in sea level and flow alterations of the Volga River, which have altered ecology of the Caspian, and to industrial pollution associated with petroleum, which has significantly contributed to the decline in the fisheries (Kasymov, 1971).]

Although most evidence indicates that chronic pollution by petroleum has not seriously affected fishery productivity along large sections of the coasts of Louisiana and the North Sea, the threat of chronic pollution on the fisheries resources of restricted sections of these coasts cannot be dismissed. Indeed, along the gulf coast of Texas, lower yields of fishery species were found in small tidal creeks receiving petroleum wastes than in similar creeks not receiving such additions (Spears, 1971). Although adult fishes may hardly suffer, nursery areas may be seriously affected as a result of smothering or poisoning of the multitudes of small organisms living in and on the top layer of the sediments on which juvenile fishes and shrimps feed. However, such effects were not associated with the *Torrey Canyon* disaster. Studies that show the effect of oil spills on nursery grounds should be conducted.

Generally, fish do not suffer directly from sinking of oil but may acquire an aberrant flavor by feeding on benthic organisms carrying oil droplets. Even if the area were an important feeding ground for fishes of commercial importance, an appreciable mortality would not occur among such fishes, but they might become tainted, which would affect the fish industry economically.

Where spills occur in spawning grounds, sinking of oil may have serious adverse effects during or shortly after spawning periods for a species like the herring, which deposits its eggs on various bottom objects. Nursery grounds that are located in shallow sheltered areas, such as the Wadden Zee or the Zealand estuaries of the Netherlands, could be especially vulnerable when large quantities of oil destroy the food organisms for the juvenile fishes and shrimps. To date, however, this has not been recorded. Tainting is not a problem in this area because the juvenile fishes mature in offshore waters.

The condition of commercial fisheries on other coasts near large oil production areas, such as those in Venezuela, the Persian Gulf, the Black Sea, and Indonesia, is not available. Until these fisheries and those from areas of heavy tanker traffic are evaluated, it would be premature to extend the results from Louisiana and the North Sea to other areas. Physical and biological environment vary so much that generalizations based on the available information could be misleading.

PERSISTENCE OF EFFECTS ON COMMUNITIES

Communities differ greatly in the time required for reestablishment following catastrophes. Planktonic communities display seasonal changes and their generation times are measured in weeks or months, while some benthic communities require years to attain maturity. Presumably persistence of catastrophic effects would be much more enduring in a long-lived benthic community than in a planktonic assemblage.

Investigations of a few selected species can be misleading if the chosen species are unusually hardy or prolific. The most intensively thorough study of a community following an oil spill has been of the West Falmouth spillage. Studies relating to the spillage are still in progress, but results indicate persistence of detectable effects for at least 2 years in the most intensely polluted areas. The West Falmouth spill affected a sedimentary bottom. Benthic communities of sediments probably develop more rapidly than some of the long-lived benthic associations of rocky bottoms. Thus, the study is measuring persistence of intermediate duration.

Continuing studies of rocky bottom benthos at the *Tampico Maru* site in Baja California showed that species numbers rose for more than a decade after the oil spillage (North, 1973) (Figure 4-1). Because the study has yielded little quantitative data, results can only be considered as indicative.

Extensive biological studies were conducted following the Santa Barbara Channel spill. This spillage caused relatively little detectable damage. Effects persisted for approximately a year (Straughan, 1972). Sedimentary, rocky bottom communities and pelagic organisms were examined. [The relatively short duration of effects may be related to the small amount of damage recorded.]

RECOVERY OF COMMUNITIES

Complete recovery means that the faunal and floral constituents that were present before the oil spill are present with their full complement of constituent age classes. Within this definition, the time interval since the West Falmouth oil spill has been too short to indicate whether recovery has been complete. Most of the original faunal elements are present at certain stations.

Two faunal constituents, which are at opposite ends of the spectrum in their response to oil-induced stress, are various species belonging to the amphipod crustacean family Ampeliscidae and the opportunistic polychaete worm *Capitella capitata,* previously mentioned.

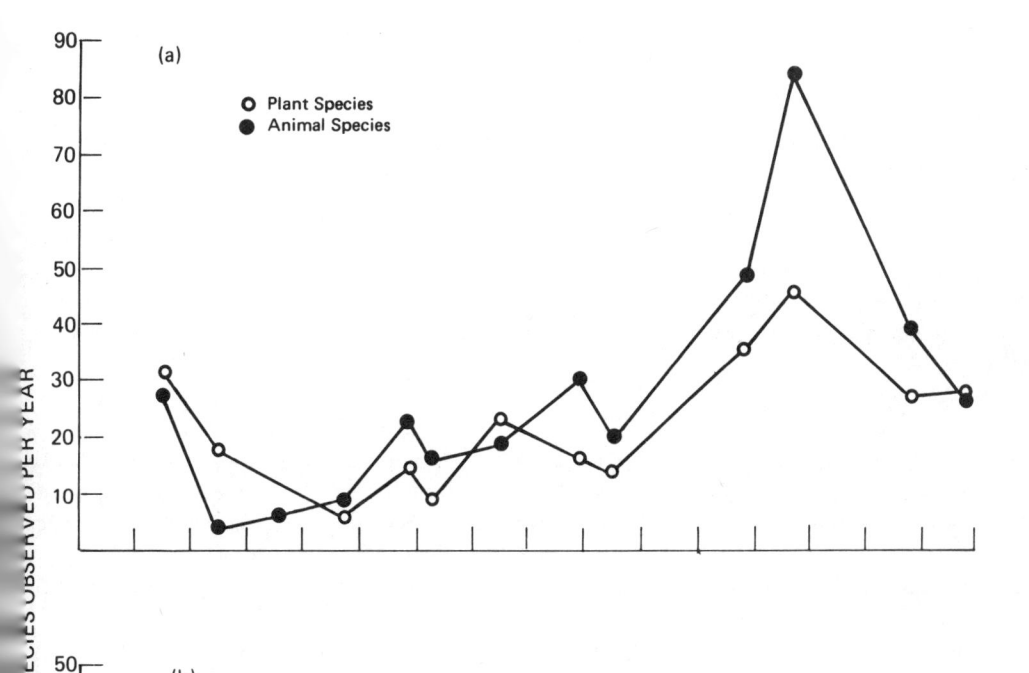

FIGURE 4-1 Variation in total numbers of plants and animal species observed (a) subtidally and (b) intertidally within a small cove at the Tampico study site. Water conditions affected efficiencies of the surveys.

Ampeliscid amphipods, although highly sensitive to even small amounts of oil, are short-lived and produce two generations each year (Sanders, 1956). The West Falmouth oil spill study shows that new recruits continuously swim into areas where ampeliscids had previously been killed or significantly depressed by the infusion of No. 2 fuel oil. If conditions remain adverse, they are killed. However, when conditions are no longer deleterious, the ampeliscids build tubes in the sediment and persist. Thus, this faunal constituent, although highly sensitive to oil, can recover completely and rapidly once the oil is degraded below the level of threshold effects.

The number of living and dead ampeliscids is arranged sequentially from highest to lowest in oil concentration. Such ordering approximates an onshore–offshore transect with highest concentrations of oil present inshore. The initial oil analyses revealed that the most distant offshore stations contained from 25 to 27 times less oil than in the heavily saturated sediments of the inshore stations (Sanders *et al.,* 1972).

Ampeliscids from the farthest offshore stations were only marginally depressed, and they completely recovered in 2 months. At stations further inshore, the impact on the ampeliscids was more severe, and complete recovery took nearly a year. Other stations located in Buzzards Bay, immediately outside of Wild Harbor, required from 18 to 28 months. However, this recovery was delayed by recycling of less degraded No. 2 fuel oil in the Wild Harbor marsh area in 1971. Thus, the temporal and spatial findings for the ampeliscids clearly indicate that the intensity of the initial impact and the time required for recovery correlated with the concentrations of No. 2 fuel oil.

The polychaete *Capitella capitata* has the ability to increase explosively in number when a habitat becomes biologically undersaturated as a result of a catastrophe. Because of its remarkably broad tolerance, *Capitella* can be present in highly stressed environments that would exclude nearly all other species.

When considering recovery in terms of human use, it should be noted that the shellfish beds in Wild Harbor River were still closed to exploitation nearly 4 years after the oil spill that severely contaminated this resource.

COMMUNITY DIVERSITY

One of the major features of biotic communities is their diversity; that is, the number of species present and their numerical composition. A high diversity means that many species are present with individuals divided relatively equally among the species; a low diversity, that few species are present with relatively marked inequality in the apportionment of individuals among the species. Usually a high diversity is interpreted as a response over an extended period to a low-stress environment or significant recovery after a stress to the high levels that may have prevailed previous to the stress.

The aftermath of any catastrophe or significant stress, including those induced by an oil spill, will result in short-term high diversity composed of opportunistic species as shown in the West Falmouth oil spill. Opportunists characteristically mature quickly, have high reproductive rates, and good dispersal abilities. Because the habitat is "biologically undersaturated" as a result of the extensive mortalities brought about by the stress, the environment is open and numerous species and individuals can invade and coexist. The animals composing these populations are almost uniformly young individuals. However, as the animals grow in size, space and food resources become limiting; thus, faunal elements must interact more intimately, decreasing diversity and density.

This type of diversity must not be confused with equilibrium diversity, which is derived from species represented by all age classes, with low reproductive rates and later maturing. The habitats are relatively saturated biologically and species' numbers and densities do not undergo marked fluctuations. Such communities tend to perpetuate themselves.

Diversity measurements do not differentiate between these two categories of diversity that have very different ecological meaning. Thus, the life history characteristics of the faunal constituents must be understood to interpret a diversity measurement correctly.

EFFECTS ON SEABIRDS

Although a wide variety of seabirds may be adversely affected from oil pollution, only auks (murres, guillemots, razorbills, puffins, etc.), penguins (though because of their geographical distribution in the Antarctic, at present only the Cape Penguin of South Africa is seriously at risk), and diving sea ducks (scoters, scaup, eiders, golden-eye, long-tailed duck) are severely damaged. These birds spend most of their lives on the surface of the sea, dive to collect their food, and are weak fliers or flightless. They dive rather than fly up in response to disturbance; if they dive on encountering floating oil or if they surface, they become completely coated with oil. Since these birds are also highly gregarious, it is possible for a small oil slick to cause very large casualties.

CHRONIC POLLUTION

Considerable stretches of the Dutch coast (later extended to parts of Belgium and France) and British coast have been routinely surveyed for the last 25 years for birds still coated with oil in nearby beach areas. For the most part, excluding heavy but localized casualties from known oil spills, the results show a fluctuating but continuing annual loss of birds from oil pollution. Most of the oil slicks that caused these casualties were small or temporary and have not been identified.

Although annual fluctuations in losses occur, there has probably been no improvement in the casualty rate in the last 25 years despite legislation and international agreement to reduce the discharge of oil wastes into the sea. The improvement in the British figures is probably because the earlier survey measured areas known to be oil polluted and the later surveys measured all areas.

The most important casualties on British coasts are auks, chiefly guillemot. They are birds most often recorded in beached bird surveys, and in most years, 80–90 percent stranded guillemots have been oiled. Divers (loons) found on the beach are nearly always oiled, but represent a small proportion of the total bird casualties. A smaller proportion (40–50 percent) of "gulls," including gannets and fulmars, are oiled and many of these, particularly Larinae, are only lightly contaminated. Mortality of these birds from oil pollution is probably relatively small.

On the Dutch and east Scottish coasts, ducks (most importantly scaup, common scoter, long-tailed duck, and eider) are the principal casualties, but a wide range of other birds are also affected by oil pollution.

Table 4-4 gives a monthly analysis of guillemot casualties on the British coast in 1970 that is also representative for other years, most other species, and for the Dutch coast and southern Baltic. Most casualties occur in winter when there are large concentrations of wintering birds feeding in these coastal areas.

Beached bird surveys on other coasts, such as the Avalon coast, Newfoundland (Tuck, 1960), Danish coast (Halt-Mortensen, 1971), have been studied for only a year and cannot be used as an indication of the mortality of birds from chronic oil pollution.

POPULATION CHANGES

A decline in breeding colonies of auks in southwest and west Britain has been evidenced. Puffin colonies of 10^4–10^6 breeding pairs in the early 1900s were reduced by two orders of magnitude by the 1960s and some colonies have disappeared altogether; guillemot and razorbill colonies have been reduced by one order of magnitude since the 1930s (Parslow, 1967). Similar trends have been observed on the Brittany coast of France (Milon, 1966). Guillemot and razorbill colonies on the Newfoundland coast have also been reported to have declined seriously, with one colony of guillemots having been reduced by 250,000 in 2 years (Tuck, 1960); razorbills, which were once numerous, are now extinct as breeders (Giles and Livingston, 1960).

Heavy and repeated losses of ducks wintering in the southern Baltic and eastern North Sea were followed by a marked fall in the number of long-tailed ducks migrating through Finland with numbers in 1958–1960 only 10 percent of those in 1937–1940 (Bergman, 1961; Lemmetyinen, 1966). Although there has been some recovery since, velvet scoters have been reported to have declined in numbers in Sweden (Lundberg, 1957).

A substantial proportion of the total population of jackass penguin, which has a restricted distribution around the coast of South Africa, has been lost in oil spills during the last decade (Westfall, 1969). Whether the size of the breeding population has been reduced as a result of spills is not known; however, it is very probable.

Auks and penguins have an extremely low replacement rate and high early mortality from natural causes. They lack the potential to replace losses caused by increased adult mortality. Oil pollution has been responsible for at least part of the decline in auk colonies in the southern part of the Atlantic. The breeding populations of puffin, guillemot, and razorbills on the Island of Rouzie on the Brittany coast were reduced by 80–88 percent in the year following the *Torrey Canyon* oil spill (Goethe, 1968). However, other factors may also be implicated. Climatic changes in the North Atlantic during the last half century may have caused a northward retreat of subarctic species, of which auks form an important group.

Diving sea ducks have a much greater reproduction

TABLE 4-4 Seasonal Distribution of Beached Guillemot Corpses on British Coasts, 1970[a]

Month	Length of Beach Surveyed (km)	Number of Corpses	Corrected Percentage for Month[b]
January	163	94	23.0
February	91	37	15.9
March	121	50	16.3
April	168	42	10.0
May	44	11	10.0
June	48	2	1.5
July	48	2	1.5
August	39	0	—
September	53	7	5.1
October	53	6	4.4
November	55	4	2.8
December	59	14	9.5

[a] From IMCO (1973).
[b] Percent of annual total, allowing for length of beach surveyed.

potential and, except for long-tailed ducks and velvet scoters, appear not to have declined despite repeated heavy losses from oil pollution. Eiders exemplify their recovery potential. A large proportion of a local diving sea duck population can be adversely affected by a single oil spill in winter; 50 percent of the Tay estuary population was lost in 1968 in the *Tank Duchess* oil spill (Greenwood and Keddie, 1968). Furthermore, since the birds return to the same nest to breed in successive years, losses must be made good by the survivors. After the *Palve* oil spill in the Kokar peninsula, in which 25–33 percent of the local population was killed, reproduction in the following year was exceptionally successful, and recovery of the population is expected to be rapid (Soikkeli and Virtanen, 1972).

REMEDIAL MEASURES

Since oil damages seabirds only when it exists as a slick, the most effective remedial measures are those that remove the oil slick as quickly as possible. In the present state of technology, this involves the use of chemical dispersant emulsifiers. In Britain, contingency plans exist for dispersing oil slicks at sea that threaten important concentrations of seabirds.

In practice, it is doubtful if preventative measures of this kind can be sufficiently effective in preventing a continuing heavy loss of seabirds from oil pollution, although they may reduce it. Rescue and cleaning of oiled birds is even less effective. The most positive conservation measure would be to improve the reproductive success of species such as auks.

NEED FOR EVALUATING BIOLOGICAL EFFECTS
OF OIL IN THE MARINE ENVIRONMENT

TYPES OF BIOLOGICAL STUDIES

Ideally, the study of the biological effects of a major oil spill should be interdisciplinary, because many environmental factors influence the intensity and duration of the impact. The immediate distribution of the oil is largely determined by winds and currents. Chemical studies are needed to evaluate the changes in composition that result from solution, evaporation, biodegradation, and weathering. The chemical analyses should evaluate the amount in the water, in the slick and in the sediments, and the changes in chemical composition that occur with time.

The biological studies should consider both plants and animals in the various habitats (intertidal and subtidal benthos, plankton, and fishes) that may be affected by the spilled oil. It would be desirable to have analyses performed before the spill; but because time and place of accidents are unpredictable, "before-the-spill" data will be rare. The next alternative is to select control areas as nearly as representative of the site of the spill, but far enough removed to avoid future contamination. The control areas would differentiate between seasonal or random changes in the populations and the direct effect of the oil spill. The first observation must be made in the contaminated area immediately after the spill; otherwise, the immediate damage may be easily overlooked, since the killed and moribund organisms may quickly disappear from the area.

A complete study has rarely been accomplished. The individual component studies have been carried out separately on different accidental spills, and since only parts have been done on each spill, most studies are not directly comparable. It is from such a montage that it is necessary to attempt to develop a coherent evaluation of the biological impact of accidental oil spills.

QUESTIONS THAT NEED ANSWERS

Serious attention has been given to the biological effects of oil spills only during the last 10–15 years. In general, the scientific evaluation of most spills is inadequate in all areas. Several critical questions must be answered before any dependable predictions can be made about potential impacts of an accidental spill, especially about the global impacts of repeated spills if they continue to increase in frequency and magnitude. Without assigning specific priorities, seven critical questions, followed by a brief explanation, are listed.

What Are the Impacts of Oil in Sediments?

Data from the Santa Barbara blowout (Kolpack, 1971) and the West Falmouth oil spill (Blumer *et al.*, 1971) document the incorporation of oil from a known source into the sediments. In more recent studies using gas chromatography, Kolpack found oil in the Santa Barbara basin sediments below 500 m of water (Kolpack, personal communication). Are these findings a general phenomenon resulting from an oil spill or blowout? What are the rates of degradation and toxicity in subtidal sediments? In view of the previous lack of awareness of this problem and the very few subtidal benthic studies addressed to the impact of oil on the bottom fauna that have been more than casual, this area of oil pollution biology deserves careful scrutiny, including re-evaluation of previous oil spills.

*What Are the Regional Effects of
Local Oil Pollution?*

The character and amount of the biological damage resulting from a local spill should be examined at regular intervals after the spill. A heavy loss of adults or of developmental stages of a particularly fecund species may be ascertained very quickly, but damage to a species with low reproductive potential may have a prolonged and widespread effect. Assessment of the significance of an oil spill, therefore, requires more than a count of immediate losses, and more consideration should be given to this differential effect on a regional basis. After the oil contamination has fallen to a low level, it is important that the effects of natural variations be considered.

What Is the "Mass Balance" of a Given Oil Spill?

Methods should be developed for determining the mass balance of an oil spill. For a given type of oil, the conditions and parameters that determine how much of the oil evaporates to the atmosphere, how much accumulates at the air–water interface, how much goes into the water column by dispersion and solution, and how much goes into the sediment must be known. To develop this information, the mechanism for transport into each of these reservoirs must be understood. Such knowledge would assist in defining which cleanup method should be employed, which ecological compartments would be most vulnerable for a particular spill, which precautions and corrective measures should be taken to minimize the impact of that spill.

What Are the Specific Toxicities of Classes of Constituents of Oil?

Predictive abilities would be greatly enhanced by laboratory-determined specific toxicities of various categories of chemical substances found in oil, provided they are related to field observations. Knowledge of toxicity mechanisms, synergistic effects, and possibilities for detoxification would also be useful.

What Is the Persistence of Oil in the Environment and the Biological Recovery from Oil Spills?

Areas affected by oil spills should be studied for long time periods to provide data on persistence of ecological effects and recovery patterns. Such studies should employ the most sophisticated analytical techniques available in both chemistry and biology.

What Is the Transport of Oil through the Food Web?

Oil spilled at sea is subject to various weathering processes—physical, chemical, and biological. Among the biological processes are bacteriological breakdown, but there are other pathways through various food chains and food webs. Via filter feeders of various description, dispersed droplets of oil and of certain components of the spilled oil may make their way to higher links in the food chain, man included, whereas oil that has become incorporated in bottom sediments may in various deposit feeders become incorporated into the bodies of series of links in various food chains, partly ending on the land.

There is insufficient information about which percentages of which components of various types of oil enter the food chains and about their ultimate fate.

How Effective Are Treatment Methods for Cleanup?

The factors and considerations that define the optimum method for cleanup of an oil spill should be better understood. For example, consideration must be given to the type of water body involved, i.e., river, lake, estuary, bay, channel, coastline water, or a part of the open ocean. Weather and oceanographic conditions—wind, temperature, currents, waves, and periods of swell—also play a role. For a given spill, a priority should be established for the relative importance of the nearshore benthic area compared with areas farther from shore: the intertidal area, the beach itself, wildlife habitats and sanctuaries, marshlands and other unique shoreline areas, living resources or marinas. An assessment of these conditions would provide a better basis for determining the proper cleanup method.

DIRECT IMPACT ON HUMANS

Oil pollution of the marine environment has two obvious direct impacts on human beings that need to be discussed. One is the degradation of aesthetic quality and its consequent effect on human physical comfort; this is caused by tar residues on beaches and shorelines. The other is the possible direct effects in human health arising from oil in the sea.

TAR AND OIL POLLUTION OF BEACHES

Perhaps the most obvious effect of marine oil pollution is the residue stranded on beaches, particularly in areas of high recreational use. This occurs not only when local acute oil spills happen but also on a worldwide chronic basis in areas far from tanker lanes and bunkering ports. The loss of aesthetic resources is substantial.

Although numerous incidents of beach pollution by acute spills and the chronic occurrence of tar on rocks and beaches of the Mediterranean coast have received attention over the past 10 years (ZoBell, 1964; Nelson-Smith, 1973), only a few quantitative or systematic studies of beach pollution by tar or other petroleum residues have been made—the East Coast of the United States (Dennis, 1959), Southern California (Ludwig and Carter, 1961), and Bermuda (Morris and Butler, 1973; Butler et al., 1973).

During 1958, Dennis (1959) found the heaviest deposits of tar on the New Jersey coast north of Atlantic City (19 g tar or oil residue on the surface of the sand in the intertidal area per meter of shoreline), on Cape Cod north of Provincetown (45 g/m), and near Chesapeake Bay (81 g/m). In most other places less than 5 g/m tar was found. Ludwig and Carter's (1961) survey showed 3–100 g/m on California beaches at the same time. More recently, the U.S. Coast Guard (1973) has measured the daily amount of tar stranded on Golden Beach, on the east coast of Florida between Miami and West Palm Beach; this ranged from less than 0.1 to 80 g/m. This year's average was 5.4 g/m. Dennis (1959) obtained in 355 days of collecting an amount corresponding to 3.5 g/m on a daily basis. Although the technique of collecting may have been somewhat different in the two cases, this is the *only* long-term quantitative comparison available in the literature, and it would appear that the average amount of tar on Golden Beach has not increased by more than a factor of 2 in the last 12 years.

There has been anecdotal evidence that beach tar on Bermuda has increased dramatically since 1967 or 1968, but quantitative surveys only began in 1971. The amount of beach tar collected ranged from 5 to 1,700 g/m, with a yearly average of approximately 190 g/m for both 1971 and 1972 (Butler et al., 1973). This is 40 times as much as the average amount found on Golden Beach, Florida (U.S. Coast Guard, 1973), during the same period. Little correlation with local wind direction was noted, and it would appear that Bermuda is acting as a collector for pelagic tar from a strip of water approximately 20 km wide. The chemical composition of beach tar (as determined by gas chromatography) is indistinguishable from that collected at sea in most parts of the Atlantic Ocean.

On high-energy sandy beaches, such as Bermuda's south shore, the self-cleaning action of high surf and offshore winds can remove virtually all the tar in one tide cycle. In rocky areas, even with high wave energy, removal by natural means may be essentially negligible. One tar globule stranded on a rock on Bermuda's south shore was sampled periodically for 16 months; similar samples were obtained from Martha's Vineyard for 13 months. Both were analyzed by gas chromatography (Blumer et al., 1973) and were found to degrade primarily by evaporation of the more volatile components. Even after a year of weathering, the basic composition of the original paraffinic crude oil was retained.

Because of the widespread occurrence of tar on the surface of the ocean, it is not surprising that other oceanic islands have beach tar comparable in amount with Bermuda. For example, a report from Zanzibar in 1972 (R. E. Morris, private communication) noted that "1 to 1.5 kg of oil wastes are recoverable by a 15 minute search over an area of 100 m² at high water mark." This implies a tar level of 10–100 g/m or one comparable with that found in Bermuda. Similar anecdotal evidence implies that the entire coast of the Mediterranean, Red Sea, and much of the remaining coast of Africa is heavily polluted with tar.

Analysis of tar lumps from Golden Beach, Florida (Attaway et al., in press), showed that they contained iron from 0.01 to 1.28 percent as Fe_2O_3, implying an anthropogenic source for beach tar because there is no iron in crude oil unless it has come in contact with tankers, bilges, and other engine wastes. Detailed gas chromatographic analysis of Bermuda beach tar (Butler et al., 1973; Blumer et al., 1973) has shown waxy paraffinic (C_{25}–C_{40}) components characteristic of crude oil sludge and occasional crystalline inclusions of essentially purely paraffinic wax. In support of this is the dramatic increase in the occurrence of beach tar on Bermuda since the closing of the Suez Canal in 1967 (Butler et al., 1973). A few samples of tar from areas near the U.S. coast have been analyzed (U.S. Coast Guard, 1973) and do not show substantially different character.

Thus, at least on the Atlantic coasts, one may infer that most beach tar is crude oil sludge, a product of that relatively small fraction of the world tanker fleet that does not use the Load on Top method.

A beach tar survey program carried out systematically in a number of geographic locations over 10 or more years could be an effective monitoring system for the type of petroleum residue most likely to result in aesthetic damage. It appears that control of such wastes is technically possible, but it remains to be seen whether international cooperation can make the elimination of beach tar a practical reality.

POSSIBLE HUMAN HEALTH EFFECTS

The possible human health effects of ocean pollution by petroleum and its by-products might be acute or chronic. Acute toxicity due to swallowing oil directly or inhalation of petroleum fumes is not relevant to this report.

For the general human population, which has limited direct contact with the ocean or its shoreline, it is reasonable to assume that if there are adverse health effects of oceanic oil pollution they would come from eating oil-contaminated seafood. The area of hazard that seems to be most worth investigating is the possible ingestion of carcinogenic compounds introduced into seafoods by oil contamination. Certain carcinogens, particularly polycyclic aromatic hydrocarbons (PAH), are known to be present in very small amounts in crude oils; therefore, minute amounts might be ingested by eating oil-contaminated seafood. However, to keep a sense of proportion, it must be kept in mind that these compounds are also assimilated from smoke of cigarettes and may be inhaled in the smoke from burning coal or petroleum, from burning refuse, from motor vehicle operations, and from the production of coke. The prototype compound benzo[a]pyrene (B-a-P) has been reported as being present in crude oil at concentrations of approximately 1 mg/kg, which when combined with the estimate of about 5 million tons of petroleum entering the ocean per year as indicated in Chapter 1, leads to an estimate of approximately 5 tons of B-a-P per year into the ocean. The 1972 National Academy of Sciences report *Particulate Polycyclic Organic Matter* estimates that considerably larger quantities of B-a-P are released to the atmosphere by heat and power generation (500 tons/year), burning refuse (600 tons/year), coke production (200 tons/year), and motor vehicles (70 tons/year), for a total of 1,400 tons/year. Thus, the amount of benzo[a]pyrene released into the sea by petroleum entering the ocean is small compared with that released into the total global environment from other sources.

However, a simple comparison between these widely differing values may not be valid as a way of determining human risks. Not only is there limited knowledge about the behavior of these compounds when they are exposed to the atmosphere or to the ocean, but there is almost no reliable information concerning the possible concentrating or diluting mechanisms for these compounds in the food chain of marine or terrestrial organisms. Furthermore, we have little knowledge of the relative effects of inhalation and ingestion of polyaromatic hydrocarbons in humans. Hence, whereas the contribution of oil spills to carcinogenic hydrocarbons in the global environment is much smaller than that added to the atmosphere by other processes, this does not suffice to dismiss the matter.

Another basis for comparison might be to estimate the concentration of B-a-P in seafood and to compare this with the amounts normally found in drinking water and in other foods. Studies (Andelman and Suess, 1970) have shown the concentration of carcinogenic PAH in groundwater and treated surface water to range between 0.001 μg/kg and 0.025 μg/kg. This may be taken as a likely range of concentration in drinking water supplies.

Carcinogenic PAH have been detected and extracted from a large variety of fresh plants, including vegetables, various grains, fruits, and edible mushrooms. B-a-P was found in lettuce at a maximum concentration of 12.8 μg/kg (Grimmer, 1966). In coconut oil, the PAH contents reached as high as 48.4 μg/kg. Water extracted from 1 kg of tea contained about 4 μg of B-a-P; 11 samples of coffee contained hardly any hydrocarbons at all (although PAH have been found in roasted coffee beans) (Borneff and Fischer, 1962).

Appreciable concentrations of carcinogenic PAH have been found in various fried, grilled, roasted, and smoked fish and meat products. Lijinsky and Shubik (1965), for example, report 8 μg/kg of dry smoked salmon; and Gorelova and Dikum (1965) found up to 10.5 μg/kg in home-smoked sausages (some B-a-P was undoubtedly produced by the smoking process).

Reported amounts of B-a-P in marine animals vary quite widely. The difficulties encountered in obtaining some of the earlier measurements suggest that the wide variability may be due as much to technique as to real differences in B-a-P content. In several species of marine animals tested after an oil spill, concentration of total hydrocarbons increased by a factor of 10 (Scarratt and Zitko, 1972). However, no direct measurements have been reported indicating an increased concentration of B-a-P after an oil spill.

Five hundred milligrams per kilogram is considered as a high hydrocarbon concentration, and such a figure has been found in contaminated shellfish after an oil spill. If we assume that this contaminating hydrocarbon retains the same B-a-P content as normal crude oil (i.e., about 1 ppm), the B-a-P content of the contaminated shellfish can be estimated at about 0.5 μg/kg. This converts to an increment of 5 μg/kg dry weight, which is within the same range as for other foods.

In general, marine organisms containing greater than 200–300 ppm of petroleum hydrocarbons become unpalatable and are unlikely to be ingested (Mackin and Hopkins, 1961). This, however, is not a reliable means of preventing ingestion. Although it is the bulk hydrocarbon that gives the bad taste, it is the incorporated traces of carcinogen that represent the possible health hazards. If the carcinogens should be concentrated in the food chain relative to the total hydrocarbons, we could be faced with high carcinogen content without any bad taste. Until further experiments are done, this possibility cannot be eliminated. This possibility exists for any carcinogen whether natural or introduced into a food chain by man's activities. With our present scanty evidence, there is no reason to treat the possible concentration of carcinogens from spilled oil as a significant threat.

To these considerations the following must be added. Knowledge of the properties of all the constituents of petroleum is not complete; therefore, there might be other dangerous materials present in petroleum that have not yet been identified. Certainly, B-a-P is only perhaps 1–20 percent of the total carcinogenic polyaromatic hydrocarbon in various environmental sources (Andelman and Suess, 1970). Furthermore, much of the information concerning the effects and metabolism of such compounds is based on experience with animals and not with direct information on or contact with humans. In short, this whole area of possible human effects from environmental chemical carcinogens is one of very considerable ignorance.

For the above-mentioned and other reasons, estimating the degree of risk from this source is extremely difficult. The mechanisms by which chemicals induce cancer are poorly understood, and those inducing birth defects are even less understood. Unfortunately, it is necessary to have some theoretical understanding of the process in order to estimate the risks due to exposure at very low concentrations since it is impossible to carry out the appropriate experiments with animals, let alone humans. We are concerned with the presence of compounds that are generally found in foods in the range of less than a few hundred parts per billion and whose effects may not be evident for several years. Thus, it is only possible to make rational public policy if we assume some shape of the dose–response curve is representative at very low doses, even if we cannot obtain sound experimental data to validate this assumption.

If these potential toxic effects exist, even in the range where the probability of the effect is very low, we must be seriously concerned because of the potential exposure of very large populations. To maintain a reasonable degree of prudence in these matters, it seems clear that we must operate under the assumption that there is no safe value, i.e., no threshold, below which complete safety can be guaranteed. On the other hand, we should attempt to evaluate the various hazards to man from compounds in his food in such a way as to minimize the total risk. It makes no sense to stop eating fish for fear of their possible carcinogenic content and replace the fish by another food source that poses an equal or even greater danger. Thus, we believe that a special effort should be made to measure the concentration of the carcinogenic PAHs in a variety of foods on a continuing basis. Although it is clear that much more information relating to possible low-level toxic effects of contaminants in all foods would be of great importance, it does not appear that our present information provides a basis for alarm about the health effects of oil spills.

Conclusions

A review of the literature (Table 4-1) shows that a limited number of documented studies exist that consider the biological, chemical, and physical acute and long-term effects of oil in the marine environment. Because most studies have been made in estuaries, little data are available concerning effects on the open ocean. However, certain generalizations about various aspects of oil in the marine environment can be made.

Whereas the concentration of petroleum hydrocarbons dissolved in water is generally low (<10 ppb) (Gordon and Prouse, in press), it was found to be much higher in sediments, ranging from 1,500 to 5,700 ppm in polluted coastal sediments (natural indigenous hydrocarbons in sediments in nearby unpolluted areas ranged from 26 to 130 ppm). On the outer coastal shelf, concentration in sediments might be as high as 20 ppm, whereas in the deep ocean 1–4 ppm was the usual concentration (Farrington and Medeiros, personal communication; Farrington and Quinn, 1973; Blumer and Sass, 1972b).

In general, where damage was severe, the oil spill was massive relative to the size of the affected area, and the spill was confined naturally or artificially to a limited area of relatively shallow water for a period of several days. Deleterious effects may have been increased by storms or heavy surf water mixed with oil and sediments in the affected area. These effects were also generally localized, ranging from a few miles to tens of miles, depending on ecological and environmental circumstances; however, for a given quantity of oil, the more localized the distribution of the spill, the greater is the mortality.

Different oils were found to have different effects, with toxicity being most pronounced for refined distillates and physical smothering most severe with viscous crude oils or Bunker C crude oil. Refined No. 2 fuel oil was among the oils having the most toxic effects. Variations in physical environment in coastal areas were also considered in determining effects; i.e., a polluted area might experience sudden and unpredictable stresses from synergistic interactions between variable environmental factors and the oil.

The amount of oil and the type of organism afflicted was also found to be important. For example, a single coating of fresh or weathered crude oil or its derivatives on certain bird species or on seeds of plants caused death, whereas marsh plants were killed only after several coatings. In general, emergent plant life was less likely to be affected than marine biota, unless the spill occurred in tropical waters where mangroves were present. Very low concentrations of the soluble fractions of kerosene interfered with searching behavior of a marine snail. Crude oil on the shells of oysters had no effects. The photosynthesis of marine phytoplankton was reported to be reduced by 100 ppb of No. 2 fuel oil. Mortality of some organisms has been found in all major spills for which studies have been published, with the pelagic diving birds being the most obvious casualties. The extent of the mortality depended on local conditions and was greatest when the releases of oil were confined to inshore areas where natural marine resources were abundant. Intertidal organisms tended to be more resistant to stress than subtidal species. In one instance, where the herbivores were reduced, the intertidal plants on which they fed increased markedly. In laboratory studies where organisms were near their limits of tolerance to temperature or salinity, pollution products caused a much greater change in metabolic rates than when the physical conditions were nearer optimum.

The recovery of polluted areas varies greatly, depending on the flushing of the polluted area, the type of the sediments on the substrata, and the degree of isolation of its ecosystems and the kinds of organism that form them. The time periods for recovery may vary from a few months to several years. In general, the initial stages of recovery are characterized by opportunistic species that are often very productive, with a much longer time required to restore the community to one that supports more long-lived species.

One characteristic of organisms composing an ecological community that may affect its stability and rate of recovery is, for example, a slow rate of reproduction or growth. Such a characteristic increases the vulnerability of a species or ecological community to damage from oil or any other pollution. Some marine birds (auks and penguins, particularly) have very slow reproductive rates, usually only one egg per year. With the normally low rate of mortality it takes about 50 years for the population to double; thus, even if oilings were widely spaced in time, they would be chronic catastrophes to auks. Such animals might never recover from a series of spills.

Marshes or estuaries, well-isolated from each other, as they are on steep coastlines such as the West Coast of the United States, provide a measure of the effects of isolation. Certain common species that live only in brackish regions of estuaries have plankton larvae. If these drifted passively in the current, they would be washed out into the open sea and lost; instead, they dive deeper after drifting toward the mouth of the estuary and are carried by the deeper currents back up to where they were spawned in the brackish regions (Bousfield, personal communication). Thus, if the estuary is an isolated one, almost all the recruitment of these organisms is from the offspring of the resident population. If this population were completely destroyed by pollution, recolonization by chance immigration from a distant estuary would probably take a very long time. The resident population of estuaries provides shelter and food for the young stages of many commercially important marine organisms (shrimp, fish, etc.).

Partly because of their isolation, the ecological communities of coastal marshes and estuaries are particularly vulnerable to the activities associated with petroleum exploration and production. The dredging to install rigs and pipelines may severely alter an estuary, and changes in the hydrology that bring about a greater incursion of higher salinity water may have severe effects on the aquatic life attuned to a given amount of salinity. For example, the increase in salinity may greatly decrease the yield of oysters per acre. In Louisiana the overall yield of oysters and shrimp has not changed much, but dredging, channelization, and other activities have so altered the marshes that the oyster industry has been forced to move into less favorable habitats, with a consequent decline in the yield per hectare since 1945. At the same time, the species composition of the shrimp catch has changed: The white shrimp declined from 96 to 50 percent of the catch, while brown shrimp increased to about 50 percent. Such changes in shrimp species are often associated with changes in the salinity of the water.

There is very little data on the effect of oil on pelagic species. Without more research, it is clearly premature to conclude anything about the effects of oil on the open ocean.

Conclusions regarding the effects of oil in the marine environment on human health are based on limited information. From our interpretation of this information, modest concern rather than alarm appears to be justified. Although it is known that petroleum contains small amounts of carcinogens and possibly small amounts of other harmful materials, the amounts of carcinogens known to be in petroleum that could be ingested by eating marine organisms is estimated to be no greater than that acquired from eating any other foods. Nonetheless, to reduce potentially harmful effects to man, all sources of carcinogens, including the large source from terrestrial activities, should be investigated and, if possible, eliminated.

The field of carcinogens and man's exposure to them needs more research. As part of this research, more studies should be performed to determine how these materials enter the ocean and, subsequently, man. Studies to detect whether there are other materials in petroleum in small quantities, such as mutagens or teratogens, are also needed because such enormous amounts of petroleum are used and handled by man. At present, the admittedly very inadequate available evidence does not make it appear that dangers of this sort from petroleum in the sea are nearly as great as other exposures to man of carcinogenic and toxic materials.

References

Andelman, J. B., and M. J. Suess. 1970. Polynuclear aromatic hydrocarbons in the water environment. Bull. W.H.O. 43(3): 479–508.

Anderson, J. W. 1973. Uptake and dipuration of specific hydrocarbons from fuel oil by the bivalves *Rangia cuneata* and *Crassostrea virginica*, pp. 690–708. Background papers, workshop on inputs, fates and effects of petroleum in the marine environment. National Academy of Sciences, Washington, D.C.

Anonymous. 1955. Pacific salmon investigations: Oil pollution studies by the Service's Seattle Biological Laboratory. Commer. Fish. Rev. 17(3):35.

Atema, J., and L. Stein. 1972. Sublethal effects of crude oil on the behavior of the American lobster. Woods Hole Oceanographic Institution Tech. Rep. 72–74. Woods Hole, Mass. Unpublished manuscript.

Attaway, D., J. R. Jadamec, and W. McGowan. Rust in floating petroleum found in the marine environment. In press.

Aubert, M., R. Charra, and G. Malara. 1969. Etude de la toxicite de produits chimiques vis-à-vis de la chaine bibliogique marine. Rév. Int. Océanogr. Méd. 13/14:45–72.

Baker, J. M. 1971. Refinery effluent. In E. B. Cowell, ed. Proceedings, Symposium on the Ecological Effects of Oil Pollution on Littoral Communities. Institute of Petroleum, London.

Baker, J. M. 1973. Biological effects of refinery effluents, pp. 715–724. In Proceedings, Joint Conference on Prevention and Control of Oil Spills. American Petroleum Institute, Washington, D.C.

Bechtel, T. J., and B. J. Copeland. 1970. Fish species diversity indices as indicators of pollution in Galveston Bay, Texas. Control Mar. Sci. Univ. Tex. 15:103–132.

Bellamy, D. F., P. H. Clarke, D. M. John, D. Jones, A. Whittick, and T. Darke. 1967. Effects of pollution from the *Torrey Canyon* on littoral and sublittoral ecosystems. Nature 216:1170–1173.

Bellan, G., D. J. Reish, and J. P. Foret. 1972. The sublethal effects of a detergent on the reproduction, development, and settlement in the polychaetous annelid *Capitella capitata*. Mar. Biol. 14:183–188.

Bergman, G. 1961. The migrating populations of long-tailed duck (*Clangula hyemalis*) and the common scoter (*Melanilta nigra*) in the spring, 1960. Suom. Riista 16:69–74.

Beynon, L. R. 1970. Oil spill dispersants. In Proceedings, Symposium on Oil Spill Prevention. Institute of Petroleum, London.

Beynon, L. R. 1971. Dealing with major oil at sea. In Water Pollution by Oil. Elsevier, London.

Blumer, M. 1969a. Oil pollution of the ocean, pp. 5–13. In D. P. Hoult, ed. Oil on the Sea. Plenum Press, New York.

Blumer, M. 1969b. Oil pollution of the ocean. Oceanus 15:3–7.

Blumer, M. 1971. Scientific aspects of the oil spill problem. Environ. Affairs 1:54–73.

Blumer, M., and J. Sass. 1972a. The West Falmouth oil spill. II. Chemistry. Woods Hole, Oceanographic Institution Tech. Rep. 72–19. Woods Hole, Mass. Unpublished manuscript.

Blumer, M., and J. Sass. 1972b. Oil pollution, persistence, and degradation of spilled fuel oil. Science 176:1120–1122.

Blumer, M., J. Sass, G. Souza, H. L. Sanders, J. F. Grassle, and G. R. Hampson. 1970a. The West Falmouth oil spill. Woods Hole Oceanographic Institution. Ref. 70–44. Woods Hole, Mass. Unpublished manuscript.

Blumer, M., G. Souza, and J. Sass. 1970b. Hydrocarbon pollution of edible shellfish by an oil spill. Mar. Biol. 5:195–202.

Blumer, M., H. L. Sanders, J. F. Grassle, and G. R. Hampson. 1971. A small oil spill. Environment 13(2):2–12.

Blumer, M., M. Erhardt, and J. H. Jones. 1973. The environmental fate of stranded crude oil. Deep Sea Res. 20:229–259.

Boesch, D. S. 1973. Biological effects of chronic oil pollution on coastal ecosystems, pp. 603–618. In Background papers, workshop on inputs, fates, effects of petroleum in the marine environment. National Academy of Sciences, Washington, D.C.

Borneff, J., and R. Fischer. 1962. Carcinogenic substances in water and soil. Part IX. Investigations on filter mud of a lake water treatment plant for PAH. Arch. Hyg. (Berlin) 146:163–167.

Brockson, R. W., and H. T. Bailey. 1973. Respiratory response of juvenile chinook salmon and striped bass exposed to benzene, a water-soluble component of crude oil, pp. 783–792. In Proceedings, Joint Conference on Prevention and Control of Oil Spills. American Petroleum Institute, Washington, D.C.

Brown, D. H. 1972. The effect of Kuwait crude oil and a solvent emulsifier on the metabolism of the marine lichen *Lichina pygmaea*. Mar. Biol. 12(4):309–315.

Burns, A., and M. Teal. 1971. Hydrocarbon incorporation into the salt marsh ecosystem from the West Falmouth oil spill. Woods Hole Oceanographic Institution. Ref. 71–69. Woods Hole, Mass. Unpublished manuscript. 23 pp.

Butler, J. N., B. F. Morris, and J. Sass. 1973. Pelagic tar from Bermuda and the Sargasso Sea. Spec. Publ. No. 10. Bermuda Biological Station. 346 pp.

California State Water Pollution Control Board. 1964. An investigation of the effects of discharged wastes on kelp. Publ. 26. Sacramento, Calif. 120 pp.

Canevari, E. P. 1971. Oil spill dispersants—Current status and future outlook, pp. 263–270. In Proceedings, Joint Conference on Prevention and Control of Oil Spills. American Petroleum Institute, Washington, D.C.

Canevari, E. P. 1973. Development of next generation chemical dispersants, pp. 231–240. In Proceedings, Joint Conference on Prevention and Control of Oil Spills. American Petroleum Institute, Washington, D.C.

Cerame-Vivas, M. J. 1968. The *Ocean Eagle* oil spill. Department of Marine Science, University of Puerto Rico, Mayaguez.

Chan, G. L. 1973. A study of the effects of the San Francisco oil spill on marine organisms, pp. 741–782. *In* Proceedings, Joint Conference on Prevention and Control of Oil Spills. American Petroleum Institute, Washington, D.C.

Cimberg, R., S. Mann, and D. Straughan. 1973. A reinvestigation of Southern California rocky intertidal beach three and one-half years after the 1969 Santa Barbara oil spill: A preliminary report, pp. 697–702. *In* Proceedings, Joint Conference on Prevention and Control of Oil Spills. American Petroleum Institute, Washington, D.C.

Clark, R. B. 1971. The biological consequences of oil pollution of the sea, pp. 53–73. *In* Water Pollution as a World Problem. The Legal, Scientific, and Political Aspects.

Clark, R. C., Jr., J. S. Finley, B. G. Patten, D. F. Stefoni, and E. E. DeNike. 1973. Interagency investigations of a persistent oil spill on the Washington coast, pp. 793–808. *In* Proceedings, Joint Conference on Prevention and Control of Oil Spills. American Petroleum Institute, Washington, D.C.

Cowell, E. B. 1969. The effects of oil pollution on salt marsh communities in Pembrokeshire and Cornwall. J. Appl. Ecol. 6:133–142.

Cowell, E. B. 1971. Some effects of oil pollution in Milford Haven, United Kingdom, pp. 231–240. *In* Proceedings, Joint Conference on Prevention and Control of Oil Spills. American Petroleum Institute, Washington, D.C.

Crapp, G. B. 1971. Zoological studies on shore communities. *In* E. B. Cowell, ed. Proceedings, Symposium on the Ecological Effects of Oil Pollution on Littoral Communities. Institute of Petroleum, London.

Darling, F. F. 1938. Bird flocks and the breeding cycle. Cambridge University Press, London.

Day, J. H., P. Cook, P. Zoutendyk, and R. Simons. 1971. The effect of oil pollution from the tanker "Wafra" on the marine fauna of the Cape Agulhas area. Zool. Afr. 6:209–219.

Dennis, J. V. 1959. Oil pollution survey of the United States Atlantic coast. Publ. 4054. American Petroleum Institute, Washington, D.C.

Dewling, R. T. 1971. Dispersant use and water quality. *In* Proceedings, Joint Conference on Prevention and Control of Oil Spills. American Petroleum Institute, Washington, D.C.

Diaz-Piferrer, M. 1962. The effects of an oil spill on the shore of Guanica, Puerto Rico, pp. 12–13. *In* Proceedings, Fourth Meeting, Associated Island Marine Labs., Curacao. University of Puerto Rico, Mayaguez.

Falk, H. L., P. Kotin, and A. Mehler. 1964. PH as carcinogens for man. Arch. Environ. Health 8:721–730.

Farrington, J., and J. G. Quinn. 1973. Petroleum hydrocarbons in Narragansett Bay. I. Survey of hydrocarbons in sediments and clams, *Mercenaria mercenaria*. Estuarine Coastal Mar. Sci. 1:71–79.

Fauchald, K. 1971. The benthic fauna in the Santa Barbara Channel following the January 1969 oil spill, pp. 149–158. *In* D. Straughan, ed. Biological and oceanographical survey of the Santa Barbara Channel Oil Spill, 1969–1970. Vol. 1. Biology and Bacteriology. Allan Hancock Foundation, University of Southern California, Los Angeles.

Foster, M., A. C. Charters, and M. Neushul. 1971a. The Santa Barbara oil spill. Part 1. Initial quantities and distribution of pollutant crude oil. Environ. Pollut. 2:97–113.

Foster, M., M. Neushul, and R. Zingmark. 1971b. The Santa Barbara oil spill. Part 2. Initial effects on intertidal and kelp bed organisms. Environ. Pollut. 2:115–134.

Gardner, G. R. 1972. Chemically induced lesions in estuarine or marine teleosts. *In* Proceedings, Symposium on Fish Pathology. Armed Forces Institute of Pathology, Washington, D.C.

Gardner, G. R., M. Barry, and G. La Roche. 1973. Analytical approach in the evaluation of biological effects. J. Fish. Res. Board Can. 35:3185–3196.

Gatteleir, C. R., J. L. Audin, P. Fusey, J. C. Lacaze, and M. L. Priou. 1973. Experimental ecosystems to measure fate of oil dispersed by surface active products, pp. 497–504. *In* Proceedings, Joint Conference on Prevention and Control of Oil Spills. American Petroleum Institute, Washington, D.C.

Giles, L. A., and J. Livingston. 1960. Oil pollution of the seas. Trans. N. Am. Wildl. Nat. Resour. Conf. 25:297–302.

Gilfillan, E. S. 1973. Effects of seawater extracts of crude oil on carbon budgets in two species of mussels, pp. 691–695. *In* Proceedings, Joint Conference on Prevention and Control of Oil Spills. American Petroleum Institute, Washington, D.C.

Goethe, F. 1968. The effects of oil pollution on populations of marine and coastal birds. Helgoländer wiss. Meeresunters. 17:370–74.

Gooding, R. M. 1968. Oil pollution on Wake Island from the tanker *R. C. Stoner*. U.S. National Marine Fisheries Service, Honolulu, Hawaii. Unpublished report.

Gordon, D. C., and N. J. Prouse. The effects of three different oils on marine phytoplankton photosynthesis. Mar. Biol. In press.

Goreloya, N. D., and P. P. Dikun. 1965. BP content of sausages and smoked fish products manufactured with utility gas or coke. Gig. Sanit. 30(7):120–122.

Greenwood, J. J. D., and J. P. F. Keddie. 1968. Birds killed by oil in the Tay estuary, March and April, 1968. Scot. Birds 5:189–196.

Grimmer, G. 1966. Carcinogenic hydrocarbons in the environment of man. Erdöl Kohle 19:578–583.

Grimmer, G., and A. Hildebrandt. 1968. Hydrocarbons in the human environment. Part VI. The content of PH in raw vegetable oils. Arch. Hyg. (Berlin) 152:255–259.

Halt-Mortensen, P. 1971. Olie-Fugle. Feltornighologen 4:186–190.

Hartung, R. 1965. Some effects of oiling on reproduction of ducks. J. Wildl. Manage. 29(4):872.

Hohn, M. H. 1959. The use of diatom populations as a measure of water quality in selected areas of Galveston and Chocolate Bay, Texas. Univ. Tex. Inst. Mar. Sci. Publ. 6:206–212.

Holmes, R. W. 1969. The Santa Barbara oil spill, pp. 15–27. *In* D. P. Hoult, ed. Oil on the Sea. Plenum Press, New York.

Hueper, W. C., and W. D. Conway. 1965. Chemical carcinogenesis and cancers. Charles Thomas, Springfield, Ill.

Inter-Governmental Maritime Consultative Organization (IMCO). 1973. The environmental and financial consequences of oil pollution from ships. Report of study No. 6 submitted by the United Kingdom. Appendix 3, The biological effects of oil pollution of the oceans. IMCO, London.

Jacobson, S. M., and D. B. Boylan. 1973. Seawater soluble fraction of kerosene: effect on chemotaxis in a marine snail, *Nassarius obsoletus*. Nature 241:213–215.

Jeffries, H. P. 1972. A stress syndrome in the hard clam, *Mercenaria mercenaria*. J. Invert. Pathol. 20:242–251.

Jones, L. G., T. Mitchell, E. K. Anderson, and W. J. North. 1969. Just how serious was the Santa Barbara oil spill? Ocean Ind. (June):53–56.

Kanter, R., D. Straughan, and W. N. Jessee. 1971. Effects of exposure to oil on *Mytilus californianus* from different localities. *In* Proceedings, Joint Conference on Prevention and Control of Oil Spills. American Petroleum Institute, Washington, D.C.

Kasymov, A. G. 1971. Industry and the productivity of the Caspian. Mar. Pollut. Bull. 2:46–49.

Kauss, P., T. C. Hutchinson, C. Soto, J. Hellebust, and M. Griffiths. 1973. The toxicity of crude oil and its components to freshwater algae, pp. 703–714. *In* Proceedings, Joint Conference on Prevention and Control of Oil Spills. American Petroleum Institute, Washington, D.C.

Ketchum, B. 1973. Oil in the marine environment, pp. 709–725. *In* Background papers, workshop on inputs, fates, and effects of petroleum in the marine environment. National Academy of Sciences, Washington, D.C.

Kittredge, J. S. 1973. Effects of water-soluble component of oil pollution on chemoreception by crabs. Fish. Bull.

Kolpack, R. L. 1971. Biological and Oceanographical Survey of the Santa Barbara Channel Oil Spill, 1969–1970. Vol. II. Physical, Chemical, and Geological Studies. Allan Hancock Foundation, University of Southern California, Los Angeles. 477 pp.

Korringa, P. 1968. Biological consequences of marine pollution with special reference to the North Sea fisheries. Helgoländer wiss. Meeresunters. 17:126–140.

Korringa, P. 1973. The ocean as final recipient of the end products of the continent's metabolism. Pollution of the oceans: Situation, consequences, and outlooks to the future, pp. 91–140. *In* Okologie und Lebensschutz in internationaler Sicht. Verlag Rombach, Freiburg.

Krebs, C. T. 1973. Qualitative observations of the marsh fiddler (*Ucca pugnax*) populations in Wild Harbor Marsh following the September, 1969, oil spill. National Academy of Sciences, Washington, D.C. Unpublished manuscript.

Kuhnhold, W. M. 1970. The influence of crude oils on fish fry. *In* Proceedings, FAO Conference on Marine Pollution and Its Effects on Living Resources and Fishing, Rome, December 1970. Food and Agriculture Organization of the United Nations, Rome.

Lacaze, J. C. 1967. Etude de la croissance d'une algue planctonique en presence d'un detergent utilise pour la destruction des nappes de petrole en mer. C.R. Acad. Sci. (Paris) 265 (Ser. D):489.

Lemmetyinen, R. 1966. Damage to waterfowl in the Baltic caused by waste oil. Suom. Riista 19:63–71.

Lijinsky, W., and P. Shubik. 1965. The detection of PAH in liquid smoke and some foods. Toxicol. Appl. Pharmacol. 7:337–343.

Ludwig, H. F., and R. Carter. 1961. Analytical characteristics of oil–tar materials on Southern California beaches. J. Water Pollut. Control Fed. 33:1123–1139.

Lundberg, S. 1957. Oversikt over Sveriges daggdjurs—Och fagelfauna 1956. Sv. Nat. Arsb. 1957:157–173.

Mackin, J. G. 1970. *In* M. C. Gillespie, ed., Oil pollution and the estuarine ecosystem. Louisiana State University, Baton Rouge. Unpublished seminar report.

Mackin, J. G. 1971. A study of the effects of oil field brine effluents on biotic communities in Texas estuaries. Report to Humble Oil and Refining Company, Houston, Tex.

Mackin, J. G., and S. H. Hopkins. 1961. Studies on oysters in relation to the oil industry. Inst. Mar. Sci. Univ. Tex. 7:1–315.

Matthews, L. H. 1939. Visual stimulation and ovulation in pigeons. Proc. R. Soc. Lond. B126:557–560.

Meijs, F., H. Schmid, L. J. Jongbloed, and H. J. Tadema. 1969. New methods for combating oil slicks, pp. 263–269. *In* Proceedings, Joint Conference on Prevention and Control of Oil Spills. American Petroleum Institute, Washington, D.C.

Milon, M. 1966. L'évolution de l'avifaune nidificatrice de la réserve Albert Chapellier (les Sept iles) de 1950–1965. Terre Vie 20:113–142.

Mironov, O. G. 1971. The effect of oil pollution on flora and fauna of the Black Sea, p. 172. *In* Proceedings, FAO Conference on Marine Pollution and Its Effects on Living Resources and Fish, Rome, December 1970, E-92. Food and Agriculture Organization of the United Nations, Rome.

Mironov, O. G. 1969. Viability of some crustacea in the seawater polluted with oil products. Zool. Zh. 68(1):1731.

Mitchell, C. T., E. A. Anderson, L. J. Jones, and W. J. North. 1970. What oil does to ecology? J. Water Pollut. Control Fed. 42(5, part 1):812–818.

Mitchell, R., S. Fogel, and I. Chet. 1972. Bacterial chemoreception: An important ecological phenomenon inhibited by hydrocarbons. Water Res. 6:1137–1140.

Morris, B. F., and J. N. Butler. 1973. Petroleum residues in the Sargasso Sea and on Bermuda beaches, pp. 521–529. *In* Proceedings, Joint Conference on the Prevention and Control of Oil Spills. American Petroleum Institute, Washington, D.C.

Moulder, D. S., and A. Varley. 1972. Bibliography of marine and estuarine oil pollution. The Laboratory of the Marine Biological Association. Plymouth, Devon, U.K.

Murphy, T. A., and L. T. McCarthy. 1970. Evaluation of the effectiveness of oil dispersing chemicals. *In* Proceedings, Industry-Government Seminar on Oil Spill Treating Agents. Am. Petrol. Inst. Publ. 40:55.

National Academy of Sciences. 1972. Particulate polycyclic organic matter. National Academy of Sciences, Washington, D.C.

Nelson-Smith, A. 1968. The effects of oil pollution and emulsifier cleansing on shore life in south-west Britain. J. Appl. Ecol. 5:97–107.

Nelson-Smith, A. 1973. Oil Pollution and Marine Ecology, Chapter 3. Plenum Press, New York.

Nicholson, N. L., and R. L. Cimberg. 1971. The Santa Barbara oil spills of 1969: A post-spill survey of the rocky intertidal, pp. 325–400. *In* D. Straughan, ed. Biological and Oceanographical Survey of the Santa Barbara Channel Oil Spills, 1969–1970. Allan Hancock Foundation, University of Southern California, Los Angeles.

North, W. J. 1973. Position paper on effects of acute oil spills, pp. 745–765. *In* Background papers, workshop on inputs, fates, and effects of petroleum in the marine environment. National Academy of Sciences, Washington, D.C.

North, W. J., M. Neushul, and K. A. Clendenning. 1969. Successive biological changes observed in a marine cove exposed to a large spillage of mineral oil, pp. 335–354. *In* Pollution Marines par les Produits Petroliers, Symposium de Monaco.

Nuzzi, R. 1973. Effects of water soluble extracts of oil on phytoplankton, pp. 809–813. *In* Proceedings, Joint Conference on Prevention and Control of Oil Spills. American Petroleum Institute, Washington, D.C.

Odum, H. T., R. P. Cuzon du Rest, R. J. Beyers, and C. Allbaugh. 1963. Diurnal metabolism, total phosphorus, able anomaly, and zooplankton diversity of abnormal ecosystems of Texas. Univ. Tex. Inst. Mar. Sci. Publ. 9:404–453.

Ottway, S. M. 1970. The comparative toxicities of crude oils. *In* E. B. Cowell, ed. Proceedings, Symposium on the Ecological Effects of Oil Pollution on Littoral Communities. Institute of Petroleum, London.

Ottway, S. M. 1972. A review of world spillages, 1960–1971. Oil Pollution Research Unit, Orielton Field Center, Pembroke, Wales.

Parslow, J. S. F. 1967. Changing status of British birds. Br. Birds 60(2–46):177–202.

Perkins, E. J. 1970. Some effects of "detergents" in the marine environment. Chem. Ind. 1:14–22.

Powell, N. A., C. S. Sayce, and D. F. Tufts. 1970. Hyperplasia in an estuarine bryozoan attributable to coal tar derivatives. J. Fish. Res. Board Can. 27(11):2095.

Reish, D. J. 1965. The effect of oil refinery wastes on benthic marine animals in Los Angeles Harbor, California, pp. 355–361. In Pollutions Marines par les Produits Petroliers, Symposium de Monaco.

Reish, D. J. 1971. Effect of pollution abatement in Los Angeles harbours. Mar. Pollut. Bull. 2:71–74.

Rice, S. D. 1973. Toxicity and avoidance tests with Prudhoe Bay oil and pink salmon fry, pp. 667–670. In Proceedings, Joint Conference on the Prevention and Control of Oil Spills. American Petroleum Institute, Washington, D.C.

Rutzler, K., and W. Sterrer. 1970. Oil pollution damage observed in tropical communities along the Atlantic seaboard of Panama. Bioscience 20:222–234.

St. Amant, L. S. 1970. Biological effects of petroleum exploration and production in coastal Louisiana. Santa Barbara Oil Symposium, pp. 335–354. Univ. Calif. Publ. Mar. Sci., Santa Barbara.

Sanders, H. L. The West Falmouth oil spill. A progress report on FWQA grant No. 15080. Woods Hole Oceanographic Institution, Woods Hole, Mass. Undated manuscript.

Sanders, H. L. 1956. Oceanography of Long Island Sound, 1952–1954. X. Biology of marine bottom communities. Bull. Bingham Oceanogr. Collect. 15:345–414.

Sanders, H. L., J. F. Grassle, and G. Hampson. 1972. The West Falmouth oil spill. I. Biology. National Technical Information Service, Springfield, Va. 49 pp.

Scarratt, D. J., and V. Zitko. 1972. Bunker C oil in sediments and benthic animals from shallow depths in Chedabucto Bay, Nova Scotia. J. Fish. Res. Board Can. 29(9):1347–1350.

Smith, J., ed. 1968. *Torrey Canyon*—Pollution and Marine Life. Report by the Plymouth Laboratory of the Marine Biological Association of the United Kingdom, London. Cambridge University Press, London. 196 pp.

Soikkeli, M., and J. Virtanen. 1972. The *Palvo* oil tanker disaster in the Finnish south-western archipelago. II. Effects of oil pollution on the eider population in the archipelagos of Kökar and Föglö, southwestern Finland. Aqua. fenn. 1972:122–128.

Spears, R. W. 1971. An evaluation of the effects of oil, oil field brine, and oil removing compounds, pp. 199–216. AIME Environmental Quality Conference. American Institute of Mining, Metallurgical, and Petroleum Engineers, Washington, D.C.

Spooner, M. 1969. Some ecological effects of marine oil pollution. Pollut. Abstr. 1(4):70–74.

St. Amant, L. S. 1970. Biological effects of petroleum exploration and production in coastal Louisiana. Santa Barbara Oil Symposium, pp. 335–354. Univ. Calif. Mar. Sci. Inst., Santa Barbara, 377 pp.

Stander, G. H., and J. A. J. Ventner. 1968. Oil pollution in South Africa. In Oil Pollution of the Sea, Proceedings of an International Conference, Rome, October 7–9, 1968, Paper No. 16b.

Steel, D. L., and B. J. Copeland. 1967. Metabolic responses of some estuarine organisms to an industrial effluent. Control. Mar. Sci. Univ. Texas 12:143–159.

Strand, J. A., W. L. Templeton, J. A. Lichatowich, and C. W. Apts. 1971. Development of toxicity test procedures for marine phytoplankton, pp. 279–286. In Proceedings, Joint Conference on Prevention and Control of Oil Spills. American Petroleum Institute, Washington, D.C.

Straughan, D., ed. 1971. Biological and Oceanographical Survey of the Santa Barbara Channel Oil Spill, 1969–1970. Vol. I. Biology and Bacteriology. Allan Hancock Foundation, University of Southern California, Los Angeles. 426 pp.

Straughan, D. 1972. Factors causing environmental changes after an oil spill. J. Pet. Tech. (March):250–254.

Struzewski, E. J., and R. T. Dewling. 1969. Chemical treatment of oil spills, pp. 217–222. In Proceedings, Joint Conference on Prevention and Control of Oil Spills. American Petroleum Institute, Washington, D.C.

Thomas, M. L. H. 1973. Effects of Bunker C oil on intertidal and lagoonal biota in Chedabucto Bay, Nova Scotia. J. Fish. Res. Board Can. 30:83–90.

Todd, J. H. 1972. An introduction to environmental ethology. Woods Hole Oceanographic Institution Ref. 72-42. Woods Hole, Mass. Unpublished manuscript.

Tuck, L. M. 1960. The Murres. Canadian Wildlife, Series 1. Can. Fish. Res. Board.

U.S. Coast Guard. 1973. Summaries of research programs by the U.S. Coast Guard on petroleum residues in the marine environment. Progress Report Abstracts. Submitted to the National Academy of Sciences.

U.S. Department of the Navy. 1970. The recovery of Bunker C fuel oil from the sunken tanker *S.S. Arrow* and concurrent measures used to control oil pollution in Chedabucto Bay, Nova Scotia, during the winter of 1970. U.S. Department of the Navy, Naval Ships Systems Command, Navships Rep. 0994–008–1010, pp. 1–97, Figs. 1–43. Navships, Washington, D.C.

University of California. 1970. Santa Barbara oil pollution, 1969. FWQA Prog. No. 15080 DZR 10/70. U.S. Government Printing Office, Washington, D.C. 42 pp.

Watson, J. A., J. P. Smith, L. C. Ehrsam, R. H. Parker, W. C. Blanton, D. E. Solomon, and C. J. Blanton. 1971. Biological assessment of a diesel spill, Anacortes, Washington. Final report, prepared by Texas Instruments for U.S. Environmental Protection Agency, Washington, D.C.

Wells, P. G. 1972. Influence of Venezuela crude oil on lobster larvae. Mar. Pollut. Bull. 3:105–106.

Westfall, A. 1969. Jackass penguins. Mar. Pollut. Bull. 14:2–7.

Wilson, K. W. 1970. The toxicity of oil-spill dispersants to the embryos and larvae of some marine fish. In Proceedings, FAO Conference on Marine Pollution and Its Effects on Living Resources and Fish, Rome, December 1970. E-92. Food and Agriculture Organization of the United Nations, Rome.

Wohlschlag, D. E., and J. N. Cameron. 1967. Assessment of low level stress on the respiratory metabolism of the pinfish (*Logodon rhomboides*). Inst. Mar. Sci. Univ. Tex. 12:160–171.

Woodin, S. A., C. F. Nyblade, and F. S. Chia. 1973. Effect of diesel oil spill on invertebrates. Mar. Pollut. Bull. 4:139–143.

ZoBell, C. E. 1964. The occurrence, effects, and fate of oil polluting the sea pp. 85–118. In Proceedings of the International Conference on Water Pollution Research, London, 1962. Pergamon Press, London.

Zafiriou, O. C. 1972. Response of *Asterias vulgaris* to chemical stimuli. Mar. Biol. 17:100–107.

5 Conclusions

The quantity of petroleum hydrocarbons entering the ocean today has been variously estimated to range from 5 to 10 million metric tons per annum (mta). Our judgment, as shown in Table 5-1, is in the lower part of this range.

The first four estimates in Table 5-1 are based on data that can be at least partially documented. The last two estimates contain major uncertainties and untested assumptions.

The river runoff input was estimated from an unpublished value of 400 ppm petroleum hydrocarbons measured in sediments deposited in the mouth of the Mississippi River and supported by analyses of other rivers of the world. However, direct measurements of petroleum hydrocarbons dissolved and dispersed in river waters as well as those carried on the water surface and by suspended particles are still very limited.

Petroleum hydrocarbon inputs from the atmosphere depend on the reaction kinetics of various compounds entering the atmosphere as well as the nature and fate of the volatile and particulate reaction products. Because very little of this information exists, our estimate was made from the known influx of petroleum hydrocarbons and our general knowledge of atmosphere residence times in global precipitation patterns.

The natural seeps input was estimated from a major extrapolation from a few known seeps. This involved estimating seeps from many areas where seeps have never been identified. No satisfactory method is available for measuring seepage rates, and our current inventory of seep areas is incomplete.

The best estimate in the table is for the input associated with transportation. It can be documented from data on tankers, terminal, and ship operations. This input represents the major source of visible accumulation of petroleum hydrocarbons both on open oceans and along coast lines.

The quantity of oil entering the oceans from transportation-related sources has been increasing every year; given future increases in production and transport, it is possible that transportation-related inputs will continue to increase despite the current interest and activity in control measures. Although the United States and a few other oil-carrying countries are adopting improved measures, such measures (e.g., Load On Top) have not been accepted as common practice by all of the major oil transporters. The capability of achieving a marked

TABLE 5-1 Petroleum Hydrocarbons in the Ocean

Input	Million Metric Tons per Annum
Transportation Tankers, dry docking, terminal operation, bilges, accidents	2.133
Coastal refineries, municipal and industrial waste	0.8
Offshore oil productions	0.08
River and urban runoff	1.9
Atmospheric fallout	0.6
Natural seeps	0.6
TOTAL	6.113

reduction in the input of oil to the sea exists, but it is heavily dependent on a much wider adoption of known control measures by all countries. The immediate need is to improve the international operation, control, and surveillance of tanker and shipping operations to minimize oil spills. Emphasis should be directed toward achieving maximum Load On Top operation by all ocean-going tankers as well as the increased use of segregated ballast tankers.

Reducing inputs to coastal waters by coastal refineries, river runoff, etc., is a much more difficult problem. Progress here will require improved control of petroleum hydrocarbon sources in municipal and industrial waste water. The control of automobile emission may reduce atmospheric fallout.

There is a need for accurate, standardized techniques for chemical analysis and for biological studies so that a more reliable analysis can be made of the ultimate fate of and biological effects of petroleum hydrocarbons. Meeting this need is very difficult due to the exceedingly complex and varied nature of petroleum as well as the wide variety of biological species and environmental conditions involved.

Conflicting reports of the biological damage following coastal oil spills can be attributed in some instances to differences in sampling procedures and analytical techniques, rather than to different environmental factors. In other instances, reports of damage to biota have not been placed in the context of normal fluctuation of the biota caused by natural environmental changes. The design of laboratory experiments to evaluate biological impairment must be such as to provide reliable data without being excessively complicated or expensive.

It is particularly important that known techniques for distinguishing between petroleum and biogenic hydrocarbons be used to determine the petroleum concentration in sediments, organisms, and water. Natural hydrocarbons are widespread so that data on total hydrocarbon content are of little value without some criteria for differentiating the petroleum hydrocarbons from the natural hydrocarbons. Unfortunately, even with present techniques this distinction cannot be made for some types of sediments and the ability to distinguish petroleum from natural hydrocarbons is less reliable at low concentrations. A reliable estimate of total hydrocarbons now in the open ocean is not possible until more sensitive diagnostic methods for determining the quantities of hydrocarbons from petroleum and biogenic sources become available.

When petroleum is spilled into the ocean, it immediately begins to undergo changes through evaporation, solution, spreading, emulsification, air–sea interchange, biological degradation and uptake, and sedimentation. The composition of petroleum and characteristics of the environment—such as temperature, concentration of bacteria and nutrients, and sea state—determine the rate at which petroleum is altered. Because the fate of diffused sources is largely unknown, it is not possible to make a material balance of the input and ultimate fate of petroleum hydrocarbons in the oceans. The fate of point sources is only partially known, namely by the accumulation of lumps, tar balls, and large mats of tarry oil residues on the open ocean and beaches along tanker routes.

The fate of most petroleum spills on the sea appears to be a combination of evaporation and decomposition in the atmosphere plus oxidation by chemical and biological means to CO_2. The heavier fraction of petroleum forms pelagic tar. The total amount of petroleum on the open sea in the form of specks and floating lumps is estimated to be less than a year's input. Some fraction of this amount eventually becomes washed up on beaches and incorporated into coastal sediments. It is this portion of spilled oil that causes most public complaints. Tar masses are appearing in increased quantity in formerly unpolluted areas such as the East Coast of Africa, the beaches of Southern France, and many islands in both the Indian and Atlantic oceans. Recent reports clearly document the quantity and nature of these tar residues in areas such as Bermuda. The fact that these tars frequently have inclusions of paraffinic wax such as that originally formed on tanker compartment walls and that they have much higher iron contents than natural petroleum is evidence that most of these materials originate from tanker washings and bilge discharges, rather than diffused sources of petroleum input or seeps.

The documentation of visible tar on beaches only accounts for a fraction of the total input into the ocean. To construct a reliable model of the fate of petroleum in the marine environment surveys over large portions of the world's oceans combined with time series data at several individual stations are needed. Data on the rate of sedimentation of petroleum residues in both open ocean and coastal areas and its incorporation into marsh and tidal flat sediments where it has considerable ecological impact are particularly important.

When oil becomes incorporated in coastal sands protected from the weathering effects of sun and oxygen, its residence time may be measured in years or decades. Unless steps are taken to reduce the input to a level that can be assimilated through natural degradation processes, we will all have to reconcile ourselves to oil-contaminated beaches.

Microorganisms capable of oxidizing petroleum constituents under the right conditions have been found in virtually all parts of the marine environment that have been examined. However, reliable information on the rates of biodegradation are not available. Both laboratory experiments and some field observations have shown that microorganisms consume the least toxic fraction of petroleum (normal alkanes) in a few days or months, depending on temperature and nutrient supply. The fraction containing aromatics and naphthenes is more toxic than the alkanes and also degrades more slowly.

Larger organisms take up hydrocarbons through the gills or from fluid passing through the gut. The quantity of petroleum hydrocarbons (excluding biogenic hydrocarbons) in the total body (wet weight) of various marine organisms reported in the literature ranges from 1 to 400 $\mu g/g$. These include organisms from clean, as well as polluted, environments. Fish and lobsters have been shown to metabolize most petroleum hydrocarbons within 2 weeks, but metabolism in lower organisms is slower and the pathways are poorly understood. There is no evidence, however, for food web magnification of petroleum hydrocarbons in marine organisms. Direct uptake of petroleum hydrocarbons from the water or sediments appears to be more important than uptake from the food chain, except in special cases. Some organisms such as mussels and oysters have been shown to eliminate most absorbed petroleum hydrocarbons when placed in clean water.

An accurate evaluation of the fate of petroleum through microbial degradation and biological uptake cannot be made until better designed and more rigorously conducted field studies are carried out. Laboratory experiments involve so many factors not encountered in the natural environment and vice versa, that few data have been useful in defining the biological fate of petroleum in the marine environment.

The most damaging, indisputable adverse effects of petroleum are the oiling and tarring of beaches, the endangering of seabird species, and the modification of benthic communities along polluted coastlines where petroleum is heavily incorporated in the sediments. The first two of these effects occur predominantly from discharges and spills of tanker and ship operations. The toxicity and smothering effect of oil caused mortality in all major spills studied, with pelagic diving birds and intertidal to subtidal benthic organisms being most affected. Mortality was greatest where oil spills were confined to inshore areas with abundant biota. The effects were generally quite localized, ranging from a few miles to tens of miles, depending on the quantity of petroleum involved.

Different petroleum products have different effects. Toxicity is greatest for refined distillates, particularly those high in aromatic hydrocarbons. Physical smothering is most severe with heavy crude oils and Bunker C fuel oil. The effects of oil in different environments may vary considerably due to synergistic interactions between oil and other environmental stresses. A single coating of fresh or weathered crude oil will cause mortality in certain bird species and plant seeds, whereas marsh plants are killed only after several coatings.

Fish do not appear to suffer from oil spills as much as seabirds and benthic organisms. Fish may acquire an oily flavor from feeding on oil-contaminated organisms, and widespread tainting of fish flesh may persist as long as significant quantities of oil are present. A long-range hazard exists for some birds such as auks and penguins because they have such slow reproductive rates that marked increases in mortality may be causing their gradual elimination.

Although our information is limited, the effect of oil contamination on human health appears not to be cause for alarm. From our calculation, we estimate that the carcinogen benzo[a]pyrene concentration on a dry weight basis arising from a high level of contamination by petroleum is comparable with that of common terrestrial foods. We, of course, do not recommend eating contaminated seafood, but in most cases, because of the taste factor, not many will be tempted to do so. It is clear that this is an area in which our knowledge is grossly inadequate and that the contamination of seafood by oil is clearly undesirable.

There are those who strongly urge the use of detergents to disperse point sources of petroleum input, such as tanker spills. This practice was not discussed in detail by the workshop, but an obvious argument in favor of detergents is that the conversion of an oil spill into a diffused and disseminated form will minimize the quantity of oil eventually reaching the beaches. Thus, the use of detergents is one way to eliminate the most visible evidence of petroleum spills. The difficulty with this practice is that we do not know what happens to the dispersed hydrocarbons. Are they truly degraded, or do they simply spread the toxic effects of oil over a larger area? Nevertheless, the use of detergents to disperse the oil at the surface where it is exposed to the weathering effects of oxygen and surface organisms is clearly preferable to the use of oleophilic sands to sink oil spills to the sea bottom. Experiments on the latter technique in the North Sea clearly resulted in oil-tainted finfish and shellfish from the area for several weeks following the experiment.

In general, much more research regarding the fates and effects of petroleum hydrocarbons in the marine environment is needed. We know that the quantity of floating tar in the open ocean and of tar along coastlines has been increasing, that major spills and localized continuous discharges of petroleum hydrocarbons have damaged various species of marine life, and that low levels of petroleum may affect the behavior patterns of certain species. Studies to date indicate that areas polluted with petroleum hydrocarbons "recover" within weeks or years (depending on local conditions and the characteristics of the petroleum); however, composition of the local biological communities may be altered. The oceans have considerable ability to purify themselves by biological and chemical actions. A basic question that remains unanswered is, "At what level of petroleum hydrocarbon input to the ocean might we find irreversible damage occurring?" The sea is an enormously complex system about which our knowledge is very imperfect. The ocean may be able to accommodate petroleum hydrocarbon inputs far above those occurring today. On the other hand, the damage level may be within an order of magnitude of present inputs to the sea. Until we can come closer to answering this basic question, it seems wisest to continue our efforts in the international control of inputs and to push forward research to reduce our current level of uncertainty.